Marine Extremes

Marine extremes, as they are conceived of in this volume, encompass environments, activities, events and impacts. Extreme environments found in and around our oceans, including the deep sea and seabed as well as the frozen polar regions, are being seriously affected by both extreme behaviours (dumping and discharge of waste, illegal fishing and piracy) and extreme events (storms, tsunamis, extreme waves and marine heatwaves). The aim of this book is to highlight the multi-disciplinary knowledge and inputs needed to address marine extremes and thereafter to explore opportunities and current challenges. Safe and healthy oceans are important for economic, recreational and cultural activities, in addition to the maintenance of ecosystem services upon which we rely. This volume gathers a unique mix of researchers working on scientific aspects of biological ecosystems and physical processes together with social scientists exploring law and governance options, community preferences, cultural values, economic aspects and criminological drivers and approaches. The multi-disciplinary feature of this book breaks down barriers that arise between disparate fields of research so that integrated solutions to ocean challenges can be found. Overall, this book argues that if we are to achieve sustainable utilisation of our oceans and blue economy goals we must better understand, and respond to, the extreme environments, activities, events and impacts.

The book is a valuable addition to the literature and will be of interest to researchers in marine science, ocean governance and natural resource economics, as well as to professionals and government officials concerned with marine policy and planning.

Erika J. Techera is a Professor in the UWA Law School and UWA Oceans Institute at The University of Western Australia.

Gundula Winter is a coastal engineer and former Research Associate at the Oceans Institute of The University of Western Australia.

Earthscan Oceans

For further details please visit the series page on the Routledge website:
www.routledge.com/books/series/ECOCE

Marine Extremes

Ocean Safety, Marine Health and the Blue Economy

Edited by Erika J. Techera and Gundula Winter

 Routledge
Taylor & Francis Group
LONDON AND NEW YORK

 from Routledge

First published 2019
by Routledge

2 Park Square, Milton Park, Abingdon, Oxfordshire OX14 4RN

52 Vanderbilt Avenue, New York, NY 10017

Routledge is an imprint of the Taylor & Francis Group, an informa business

First issued in paperback 2020

British Library Cataloguing-in-Publication Data
A catalogue record for this book is available from the British Library

Library of Congress Cataloging-in-Publication Data
Names: Techera, Erika J., editor. | Winter, Gundula, editor.
Title: Marine extremes : ocean safety, marine health and the blue economy /
edited by Erika Techera and Gundula Winter.
Description: Abingdon, Oxon ; New York, NY : Routledge, 2019. |
Series: Earthscan oceans | Includes bibliographical references and index.
Identifiers: LCCN 2018045628 (print) | LCCN 2018057491 (ebook) |
ISBN 9780429491023 (eBook) | ISBN 9781138590441 (hardback) |
ISBN 9780429491023 (ebk)
Subjects: LCSH: Marine ecosystem management. | Marine ecosystem
health. | Coastal zone management. | Marine pollution. | Marine
resources–Economic aspects.
Classification: LCC QH541.5.S3 (ebook) | LCC QH541.5.S3 M2796
2019 (print) |
DDC 578.77–dc23
LC record available at https://lccn.loc.gov/2018045628v

ISBN: 978-1-138-59044-1 (hbk)
ISBN: 978-0-367-66276-9 (pbk)

Typeset in Bembo
by Wearset Ltd, Boldon, Tyne and Wear

Contents

Figures

Tables

Contributors

Karin R. Bryan is Professor in Earth Sciences at the University of Waikato, New Zealand. Her research on hydrodynamics and sediment movement in nearshore and estuarine environments has been published in 122 peer-reviewed papers and 3 book chapters, and underpins natural hazard research in New Zealand. Her recent work, both locally and in the Mekong Delta, Vietnam, explores the interactions between mangroves and coastal currents, and their potential impact on coastal flooding.

Michael P. Burton is an Associate Professor in The University of Western Australia School of Agriculture and Environment. He is an environmental and resource economist, working largely in the area of non-market valuation, applied to both marine and terrestrial environments.

Belinda L. Cannell is a Research Fellow in the Oceans Institute and the School of Biological Sciences at The University of Western Australia, and a Research Associate in the School of Veterinary and Life Sciences at Murdoch University, Western Australia. She is a conservation ecologist specialising in seabird research, particularly the ecology of Little Penguins. She has been involved in a long-term programme monitoring Little Penguins in Western Australia for more than 20 years.

Lucille Chapuis is a Research Associate within the Neuroecology Group at the Oceans Graduate School at The University of Western Australia and a member of the UWA Oceans Institute. Her research focuses on fish bioacoustics and hearing systems, with a particular interest in elasmobranchs and deep-sea fishes. She also studies the effects of anthropogenic noise on marine fauna.

Shaun P. Collin is a Professor in the Oceans Graduate School and the Oceans Institute at The University of Western Australia. His research focuses on the neural basis of the behaviour of animals from diverse environments including the deep sea. He investigates how animals detect and process sensory information such as odours, light, electric fields and sound from natural and anthropogenic sources and how these environmental cues impact behaviour. He has published over 260 papers and books.

Josie Crawshaw is an environmental scientist at the Bay of Plenty Regional Council in New Zealand. Her research has focussed on the process of denitrification in coastal lagoons and estuaries, investigating the spatial and seasonal drivers of agricultural nitrogen removal.

Matthew W. Fraser is Research Fellow in the Oceans Institute and School of Biological Sciences at The University of Western Australia. He is a marine ecologist with a research focus on understanding interactions between marine primary producers and their surrounding environments. He is also interested in understanding the bottom-up impacts of declines in marine primary producers related to climate change. He uses a range of molecular methods to investigate these interactions and to help improve the conservation of coastal marine ecosystems.

Anas Ghadouani is Professor and Programme Chair of Environmental Engineering at the University of Western Australia and a member of UWA's Oceans Institute. His research focuses on biophysical processes in water and wastewater engineering, ecological engineering and multidisciplinary research in water sensitive cities. He is editor of *Hydrology and Earth System Sciences*, *Water* (Switzerland) and *Frontiers in Marine Science: Coastal Ocean Processes*.

Marco Ghisalberti is a Senior Lecturer in the Department of Infrastructure Engineering at the University of Melbourne. His research focuses on flow, transport and biophysical coupling in aquatic ecosystems in both marine and freshwater environments. He is on the editorial board of the journals *Environmental Fluid Mechanics* and the *Journal of Ecohydraulics*.

Jeff Hansen is a Senior Lecturer in the School of Earth Sciences and Oceans Institute at The University of Western Australia. His research focuses on understanding physical processes in coastal environments and linking these to evolution of the coast, ocean observing systems and marine renewable energy. Recent projects include exploring coastal morphodynamics onshore of fringing reefs and the impacts of wave energy converters on the incident wave field and onshore coastal processes.

Christopher D. Hepburn is an Associate Professor at the University of Otago, New Zealand. He has a diverse range of interests, primarily focussing on cold-temperate kelp forest ecosystems and the values they provide. Through this work he has developed strong partnerships with local communities and helped them to face local issues such as pollution, habitat destruction, over-fishing and environmental change by providing strongly applied science with a local focus.

Yasha Hetzel is a Research Associate within the Coastal Oceanography Group at the Oceans Graduate School at The University of Western Australia and a member of the UWA Oceans Institute. His research on ocean circulation and sea level processes aims to solve environmental problems and alleviate hazards to coastal communities. In particular, he is interested in how extreme climate and weather events impact the nearshore marine environment, and how moderate events can combine to cause extremes.

Matthew R. Hipsey is an Associate Professor of Environmental Science within the School of Agriculture and Environment and the Oceans Institute at The University of Western Australia. His research focuses on understanding the interactions of physical, chemical and biological processes in aquatic systems and identifying mitigation strategies for improving water quality. His expertise is in developing modelling platforms for exploring how coastal environments respond to land use and climate change, including the effects of extreme events.

Peisheng Huang is a Research Fellow in The University of Western Australia School of Agriculture and Environment and a member of the UWA Oceans Institute. His research focuses on the hydrodynamics and water quality modelling of coastal and inland waters with particular interest in the application of numerical models on water quality restoration and real-time simulations. He also studies greenhouse gas dynamics and emissions from estuaries.

Alizée P. Lehoux is a postdoctoral researcher in the Department of Earth Sciences, University of Uppsala, Sweden. Her research focuses on various remediation techniques for contaminated soil and sediment with an emphasis on the understanding of transport and adsorption mechanisms of contaminants in porous media. Currently, she is working on the possibility of remediating fiberbanks by in-situ capping in the Baltic Sea.

Jade Lindley is a criminologist within The University of Western Australia Law School. Her research extends to international law and regulation and governance of organised crime, such as illegal fishing, maritime piracy and illicit trafficking. She is particularly interested in the Indian Ocean. She is also part of the UWA Oceans Institute leadership team for the research theme, Maritime Security, Safety and Defence. She has previously worked in research within both state and federal Australian government as well as for international organisations.

Sarah McCulloch is an environmental engineer working for Cardno in infrastructure and environmental services. She is passionate about sustainability and water resources in general. During her master's degree at The University of Western Australia, she conducted detailed experimental research on the fate and transport of microplastics and microfibres through the wastewater treatment process. She is interested in how turbulence and hydrodynamic modelling could help predict the transport of pollutants in the environment.

Nico K. Michiels is Professor in Animal Evolutionary Ecology in the Department of Biology at the University of Tübingen, Germany. His research focuses on ambient light manipulation by diurnal fish, in particular the possibility that they actively redirect downwelling light to enhance visual detection of cryptic prey or predators.

Ryan P. Mulligan is an Associate Professor in the Department of Civil Engineering at Queen's University, Canada. His research encompasses coastal engineering and physical oceanography, and particularly focuses on the impacts of extreme events on the nearshore coastal environment. These include the impacts of storms such as tropical cyclones on beaches and estuaries, and the generation of tsunamis by landslides in mountainous coastal regions.

Carolyn E. Oldham is a Professor in the School of Engineering and the UWA Oceans Institute at The University of Western Australia. Her research focuses on integrating approaches and frameworks across multiple disciplines, and as well as maintaining her core interest in the interactions between contaminant transport and biogeochemical transformation, with a focus on patchiness and connectivity dynamics at system, local and micro scales.

Chari B. Pattiaratchi is a Professor of Coastal Oceanography, UWA Oceans Institute at The University of Western Australia. His research explores coastal ocean physical processes and their influence on climatic, biological and geological processes in estuaries, the nearshore (beach) zone and the continental shelf region, utilising field measurements, remote sensing and computer modelling. He leads the Australian National Facility for Ocean Gliders at UWA.

Andrew W. M. Pomeroy is a Research Associate in the Oceans Graduate School at The University of Western Australia, a member of the UWA Oceans Institute and affiliated to The Australian Institute for Marine Science. He researches physical processes in coastal environments with an emphasis on how these processes transport material and are affected by aquaculture production. He is also a member of the National Committee for Coastal and Ocean Engineering (Australia) and a Fulbright Scholar.

William J. Rayment is a Senior Lecturer in the Department of Marine Science at the University of Otago, New Zealand. His research is focussed on the ecology and conservation of marine megafauna, particularly cetaceans. He is a council member of the New Zealand Marine Sciences Society and a trustee of the New Zealand Whale & Dolphin Trust.

Abbie A. Rogers is a Research Fellow in the Centre for Environmental Economics and Policy at The University of Western Australia. She works extensively in the field of non-market valuation. Her primary research interests are in measuring community values and preferences for environmental conservation and management, with a focus on marine and coastal environments. A feature of her research is the applied relevance of non-market values for policy and decision making.

Candida Savage is a Senior Lecturer in the Department of Marine Science at the University of Otago, New Zealand, and an Honorary Research Associate at the Marine Research Institute and Department of Biological Science, University of Cape Town, South Africa. She is a marine ecologist with research interests in nutrient cycling and fluxes of organic matter across ecosystems, food webs and benthic community structure and functioning under multiple stressors.

Verena Schoepf is a Research Fellow in the School of Earth Sciences and UWA Oceans Institute at the University of Western Australia. She also serves as Programme Co-Leader in the ARC Centre of Excellence for Coral Reef Studies. She researches the impacts of climate change on tropical coral reefs, and has published over 20 papers.

Ana M. M. Sequeira is an Australian Research Council Discovery Early Career Researcher Award (ARC DECRA) Research Fellow in the UWA Oceans Institute and School of Biological Sciences at The University of Western Australia. She is interested in the development of models to assist understanding of the marine environment with a strong emphasis in supporting marine spatial planning and conservation. She has published over 20 papers and 1 book review and has recently been involved in the European-funded project *Aquaspace – Making space for aquaculture.*

Ian Snowball is Professor of Quaternary Geology and director of the research programme Natural Resources and Sustainable Development, which is part of the Department of Earth Sciences at Uppsala University, Sweden. Ian has published over 100 papers, book chapters and popular science texts on geomagnetism, environmental change and paleoclimatology. His current research focus is the characterisation and remediation of contaminated sediments.

Erika J. Techera is a Professor in the UWA Law School and UWA Oceans Institute at the University of Western Australia. She researches international and comparative environmental law issues, predominantly in relation to oceans. Her current projects explore Indo-Pacific legal frameworks for marine conservation, fisheries management and maritime heritage. She is member of the International Union for Conservation of Nature (IUCN) Commission on Environmental Law and World Commission on Protected Areas, has consulted for the United Nations Food and Agriculture Organization (FAO) and the United Nations Environment Programme (UNEP), and is a former barrister.

Paul G. Thomson is a biological oceanographer and Senior Research Engineer in the Oceans Graduate School and the UWA Oceans Institute at The University of Western Australia. His research interests include the ecology of marine microbes (phytoplankton, protozoa and heterotrophic prokaryotes) and how marine extremes and climate change affect their distribution and abundance. He is currently employed by the underwater glider facility of the Integrated Marine Observing System (IMOS) at UWA where he works with glider bio-optical and other IMOS data.

D. G. Webster is an Associate Professor in the Environmental Studies Program at Dartmouth College, New Hampshire, US, researching the complex dynamics of large-scale social–ecological systems (SESs). She has written two books on global fisheries governance and is leading research projects on slavery in global fisheries, the impacts of risk perception on management of harmful algal blooms and responses to climate-related extreme events on coral reefs. She also leads the Earth Systems Governance task force on Oceans Governance (www.earthsystemgovernance.org/oceans/).

Gundula Winter is a coastal engineer and former Research Associate in the Oceans Institute at The University of Western Australia, and a member of the ARC Centre of Excellence for Coral Reef Studies. Her research enhances our understanding of the physical processes in coastal waters to guide coastal management in a changing climate. Currently, Gundula works as a researcher and consultant on real-time flood forecasting and coastal risk reduction at the applied research institute Deltares in the Netherlands.

Preface

The aim of this book is to highlight the challenges and opportunities presented by marine extremes such as storms and marine heatwaves, as well as the extreme consequences of human activities such as pollution, over-exploitation and criminal activity. Safe and healthy oceans are important for economic, recreational and cultural activities, in addition to biodiversity conservation. However, marine extremes are having noticeable impacts on the world's oceans, coastlines and the actors dependent on them. These impacts are expected to intensify under current climate change and growth projections.

Marine Extremes draws upon principles from the physical, economic and social sciences to assess the impacts of marine extremes on communities, infrastructure and marine ecosystems. The book reviews forecasting methods for predicting marine extremes, assesses monitoring options for gauging the impact of marine extremes and evaluates management and regulatory measures to mitigate their impacts. Finally, this book not only highlights the threats posed by marine extremes, but also the emerging opportunities – for example, through technology, aquaculture and tourism.

Erika J. Techera and Gundula Winter

Acknowledgements

This book has been an interesting and exciting journey for us. As with all books it is the product of much hard work, not only by the co-editors but all of the contributors too. We believe strongly that multi-disciplinary research is critical in ensuring the health of our oceans and the sustainability of human activities. Combining multiple fields of work is a significant challenge; yet we hope that this book demonstrates that it can be done.

We acknowledge the support of The University of Western Australia for providing funding and assistance to host the 'Oceans and Blue Economy' workshop. We thank all of the Matariki Network of Universities members for allowing their researchers to participate in the workshop which led to the development of this book. We are grateful for the time, expertise and input of all the presenters at that workshop which allowed us to develop many of the ideas in this volume. We thank the independent peer reviewers whose critical feedback helped us to refine the chapters.

We also acknowledge the support of Tim Hardwick and Amy Johnston at Earthscan for encouraging the publication of this exciting volume.

Finally, our thanks go to both our families who have learnt a great deal more than they intended about 'marine extremes'.

Abbreviations

AEP	Annual Exceedance Probability
AIS	automatic identification systems
AODN	Australian Ocean Data Network portal
ARI	Average Recurrence Intervals
ASEAN	Association of Southeast Asian Nations
atm	atmosphere
AU	African Union
AUD	Australian dollar
AZA	allocated zones for aquaculture
CHRMAP	Coastal Hazard Risk Management and Adaptation Planning
CPA	Customary Protection Area
CSIRO	Commonwealth Scientific and Industrial Research Organisation
dia.	diameter
DSL	deep scattering layer
DVM	diel vertical migrations
EAA	Ecosystem Approach to Aquaculture
EEZ	exclusive economic zone
EJF	Environmental Justice Foundation
ENSO	El Niño Southern Oscillation
EOV	essential ocean variable
EU	European Union
FAO	Food and Agriculture Organization
FIO	Faecal Indicator Organisms
GDP	gross domestic product
GIS	geographic information system
GL	gigalitre
GOOS	Global Ocean Observing System
H+	hydrogen ions
ha	hectare
HABs	harmful algae blooms
ICSU	International Council for Science
ICZM	Integrated Coastal Zone Management
ILO	International Labour Organization

IMO   International Maritime Organization
IMOS   Integrated Marine Observing System
IOC   Indian Ocean Commission (Chapter 11)
IOC   Intergovernmental Oceanographic Commission (Chapter 6)
IORA   Indian Ocean Rim Association
IOTC   Indian Ocean Tuna Commission
IPCC   Intergovernmental Panel on Climate Change
IPOA   International Plan of Action
ISA   International Seabed Authority
IUCN   International Union for Conservation of Nature
IUU   illegal, unreported and unregulated
MPA   marine protected area
m/s   metres per second
MSC   Marine Stewardship Council
MSL   mean sea level
N   nitrogen
N_2 dinitrogen gas
N_2O nitrous oxide
NAO   North Atlantic Oscillation
NEAFC   North East Atlantic Fisheries Commission
NGO   non-governmental organisation
nm   nanometer
NZ$ New Zealand dollar
OA   ocean acidification
O(cm)   order of centimetres
OECD   Organisation for Economic Co-operation and Development
POPs   persistent organic pollutants
QMRA   quantitative microbial risk assessment
RCP   Representative Concentration Pathway
RFMO   regional fishery management organisation
SCEVO   Swan–Canning Estuary Visual Observatory
SDG   Sustainable Development Goal
SGD   Submarine Groundwater Discharge
SOFAR   sound fixing and ranging
sp.   species
SST   sea surface temperature
SWWA   South West Western Australia
TC   tropical cyclone
tg   teragram
UNCLOS   United Nations Convention on the Law of the Sea
UNEP   United Nations Environment Programme
UNISDR   United Nations Office for Disaster Risk Reduction
USD/US$ United States of America dollar
VMS   vessel monitoring systems
WA   Western Australia

WCPFC	Western Central Pacific Fisheries Commission
WMO	World Meteorological Organisation
WSUD	water sensitive urban design
WWTP	waste-water treatment plants

Part I
Introduction

Part 1

Introduction

1 Introduction to marine extremes

Erika J. Techera and Gundula Winter

Introduction

Earth is indeed a blue planet with over 70 per cent of the surface covered by water. Life began in the oceans and the marine environment continues to provide a range of ecosystem goods and services without which we could not survive. From the oceans come food and other resources for sustenance, livelihoods and trade. The oceans provide oxygen and absorb carbon dioxide, as well as regulating climate and weather. We utilise marine areas for exploration, shipping and transport as well as recreation, cultural activities and tourism. The marine environment, and the living and non-living resources it holds, are therefore of considerable importance to us. Nevertheless, our activities, largely since industrialisation, have degraded our oceans over time. We continue to exploit the oceans and marine resources despite warnings that the oceans are not unlimited areas to plunder from and dump waste within. Inshore, coasts and estuaries at the land–water interface are more familiar to us, yet many ocean areas remain little-known, hostile environments. The oceans provide a range of natural extremes: deep sea trenches extending further down than the tallest mountains rise on land; temperature ranges from freezing to tropical waters; and extreme events including marine heatwaves, storms, tsunamis and waves higher than multi-storey buildings. The environmental diversity has led to the evolution of unique species and ecosystems. Concurrently, extreme events have negative impacts on human enterprises, coastal communities and natural habitats. It is therefore timely to explore what we know about our oceans and our impacts upon them, as well as future predictions, given the changing environment.

This volume explores 'marine extremes' in the context of growing concern about over-fishing, marine debris and climate change, as well as a simultaneously increasing interest in wealth from the oceans and the concept of the blue economy. Marine extremes, as they are conceived of in this volume, encompass environments, activities, events and impacts. Extreme environments found in and around our oceans include the deep sea and seabed as well as the frozen polar regions – alien areas which we have been unable to explore until recently. They may hold valuable resources yet to be exploited, including supplying commodities found on the seabed and new materials that can be developed through the application of biotechnology. Activities in our oceans can have extreme consequences or involve extreme behaviours in themselves. Accidental oil spills can have acute catastrophic effects, but more cumulative actions – including land-based marine pollution – can have globally significant consequences such as the emerging problem of marine plastic debris. These two challenges arise from lawful activities, but there is little doubt that extreme behaviours in the form of deliberate dumping of hazardous waste, illegal fishing and piracy not only impact the marine environment but also legitimate enterprises and operators. Extreme events in the oceans and coastal zones include storms, tsunamis, extreme waves and marine heatwaves. Some of these can be modelled and predicted, resulting in advances in warnings and protection measures. They are, however, exacerbated by climate change, perhaps the most pressing current issue, caused by cumulative global emissions and shifting the baselines of marine extremes. The extreme effects of climate change – increased extreme weather events, rising sea levels and acidification, as well as elevated intensity and frequency of marine heatwaves – will have devastating consequences for marine species and ecosystems, coastal areas, communities and economic enterprises.

These concerns led us to develop a research theme within the Matariki Network of Universities, entitled 'Oceans and the Blue Economy'. The Matariki Network brings together academic experts from the USA, Canada, Europe, Australia and New Zealand focussed on research excellence through collaboration (https://matarikinetwork.org/). The 'Oceans and the Blue Economy' research theme is aimed at identifying the ways in which we can safeguard and build resilience in marine environments as well as achieve blue economy goals. While there is considerable potential for extracting further wealth from the oceans – through shipping and transport, fisheries, energy and tourism – utilisation of the ocean and marine resources must be sustainably managed and balanced with environmental conservation and socio-cultural needs. This research theme seeks to discover innovative solutions including cooperation between different sectors involved in ocean exploitation and collaboration across disciplines working on marine challenges. As a starting point, a workshop was held in December 2017 to gather experts from a variety of fields and countries to explore: the challenges facing the oceans; current knowledge and developments; and the multi-sectoral and multi-disciplinary research needed to maintain healthy oceans and achieve blue economy goals. The workshop was divided into three sub-areas – wealth from the oceans, safe oceans and healthy oceans – and involved 24 presentations and over 70 researchers. This volume is the product of that workshop and the first publication output of the 'Oceans and the Blue Economy' theme.

The volume gathers a unique mix of researchers working on scientific aspects of biological ecosystems and physical processes together with social scientists exploring law and governance options, community preferences, cultural values, economic aspects and criminological drivers and approaches. The multi-disciplinary feature of this book seeks to break down some of the barriers that arise between disparate fields of research and endeavour. If integrated solutions to ocean challenges are to be found, misunderstanding and miscommunication between fields must be addressed and synergistic outcomes need to be achieved through collaboration, to which this book contributes.

Three key threads flow through the volume: (1) understanding our marine environment and predicting future changes to it; (2) analysing the impacts of marine extremes (anthropogenic impacts on the oceans and the effects of marine extremes upon human activities, coastal communities and infrastructure); and (3) responses to the challenges of marine extremes, which are drawn from both successful exemplars and the implementation of novel, innovative ideas. Contributors include researchers who explore issues from a variety of perspectives: theoretical and practical, focussed on understanding processes and problems, filling knowledge gaps, as well as presenting possible solutions, including those grounded in top-down and bottom-up, government and community-based approaches. Another key feature throughout the volume is the use of case studies. In some examples, the case studies are quite particular and focussed on a localised area; in other chapters they cover broader regional areas and global contexts. These case studies demonstrate specific outcomes of science and social science in action, and in some situations they can highlight scalable or transposable best practice. Together this combination of research-led analysis and case studies contributes to the translation of science and social science into beneficial ocean outcomes.

Exploring marine extremes

The aim of this book is to highlight the multi-disciplinary knowledge and inputs needed to address marine extremes and thereafter to explore opportunities and current challenges. Safe and healthy oceans are important for economic, recreational and cultural activities, in addition to the maintenance of ecosystem services upon which we rely. However, marine extremes are noticeably impacting on the world's oceans and coastlines as well as the stakeholders dependent on them. These impacts are expected to intensify under current climate change and blue growth projections, and therefore new knowledge and innovative solutions are needed.

The chapters in this book have been arranged to build a multi-faceted picture of marine extremes: environments, events and activities. The topics included cover offshore, coastal and estuarine environments at different depths and temperature ranges. The impacts of localised and global activities and events on humans, communities, individual species and ecosystems are explored. In doing so, this volume seeks to build a comprehensive understanding of the challenges of marine extremes and the potential solutions to address them.

The book commences with a consideration of the different bodies of knowledge and approaches needed in understanding the problems and designing solutions. In Chapter 2, 'The Science, social science and governance of extremes', Techera and Winter highlight the range of extreme environments, activities and events associated with the ocean as well as the ways in which they impact upon, and are impacted by, people. The authors succinctly set out the fundamental science and social science disciplines involved in understanding, working within and utilising marine extremes, as well as the concepts and approaches that must be integrated in order to mitigate negative impacts. Key concepts such as ecosystem-based management, the blue economy and oceans governance are explained with common threads drawn out to set the scene for the chapters that follow. The chapter concludes by examining how all these bodies of knowledge are essential to inform our efforts to ensure human safety, marine health and sustainable use of the oceans.

Part II focuses on safe oceans; how extreme marine events impact upon human safety at sea and in coastal areas and the ways in which these challenges can be addressed. The chapters traverse a range of relevant areas including the fundamentals of coastal processes, community values and preferences for interventions, as well as nature-based solutions to marine extremes.

In Chapter 3, 'Coastal processes, extreme events and forecasting', Winter, Hetzel, Peisheng, Hipsey, Mulligan and Hansen explain the different oceanographic processes that are in play in coastal and estuarine environments. It is critical to understand these processes if we are to model and predict the impacts of extreme events on shores, reefs, lagoons and estuaries as well as coastal communities and infrastructure. The authors adeptly emphasise how modelling and forecasting is of critical importance to governments tasked with implementing protection measures and ensuring public safety. They explore key challenges with reference to case study examples and conclude by highlighting how new research is advancing the field.

Chapter 4, 'Community values and preferences for coastal hazard interventions' by Rogers and Burton emphasises that effective management of coastal assets requires community input, compliance and satisfaction. The authors explain, clearly and effectively, why community values are such an integral component of coastal hazard management and illustrate approaches through which these values can be measured and integrated into decision making. They present an example of how to quantify community values for coastal assets in monetary-equivalent terms, which enables social and environmental values to be traded off with the economic costs of coastal hazard management.

The final chapter in Part II, Chapter 5, addresses 'Nature-based solutions to mitigate extreme coastal impacts'. Winter, Bryan and Ghisalberti outline how coastal areas and communities can be protected from the impacts of coastal hazards by either engineered structures or natural systems. They analyse the benefits and shortcomings of each in terms of effective coastal protection and maintenance of coastal amenity. The focus of the chapter is on natural ecosystems such as mangroves, seagrass, kelp, saltmarshes and reefs that are adaptive and can provide coastal protection services. The authors emphasise that it is important for decision-makers to understand these nature-based options and how they can protect the shoreline during extreme events without impacting upon community values.

Part III highlights how extreme events, exacerbated by climate change as well as extreme pollution, affect marine health. The chapters cover the fundamentals of monitoring to better understand the health of our oceans, anthropogenic pollution and marine heatwaves as examples of damaging impacts as well as the importance of local management.

In Chapter 6, 'Monitoring ocean and estuary health', Thomson, Cannell, Ghadouani, Fraser and Rayment, begin by stressing our reliance upon marine ecosystem goods and services, and also how we impact upon them through both land-based and oceanic activities. They highlight the critical importance of understanding how our oceans are changing, and the global push to monitor ocean health over the past 25 years. The authors explore ocean monitoring efforts globally and at the national level through the use of long term observations of physical ocean properties as well as of short and long term biological indicators of change.

Chapter 7 examines anthropogenic impacts of 'Pollution from land-based sources'. Snowball, Lehoux, Crawshaw, Savage, Hipsey, Ghadouani, McCulloch and Oldham explore the diversity of human impacts on our oceans through the lens of terrestrial pollutants. The authors focus on case studies in Europe, New Zealand and Australia involving metal and persistent organic pollutants, agricultural and urban run-off, as well as the emerging challenges of microplastics and microfibres. The chapter highlights not only the diversity and extent of the problems and impacts but also the responses in terms of research and mitigation efforts.

Chapter 8 examines the specific ocean health issue of the 'Impacts of marine heatwaves'. Cannell, Thomson, Schoepf, Pattiaratchi and Fraser explore the different impacts of marine heatwaves and how climate change scenarios indicate increasing ocean temperature, as well as heat content, which will lead to the more frequent occurrence of marine heatwaves. This chapter explores the impacts of the increased frequency and intensity of marine heatwaves in terms of species mortality, changes in the timing of life-cycles, range extensions and contractions. A variety of ecological communities is analysed through a comprehensive case study examining impacts on plankton, reefs, seagrasses, macroalgae, penguins, fish and marine pests. The chapter draws attention to the need for continued research to support enhanced understanding of the effects of marine heatwaves so that appropriate action can be taken to mitigate impacts upon their environment, communities and operations.

The final chapter in Part III, by Hepburn (Chapter 9), entitled 'Local, community-led interventions to address global-scale problems and environmental extremes in coastal ecosystems', focuses on the challenge of effective local responses to global pressures. Through an inspirational 25-year case study, Hepburn explores community-driven management in action. The author draws attention to the importance of empowering local communities, recognising indigenous knowledge and respecting cultural values. The findings confirm that effective local management of ecosystems can ensure benefits now and also build resilience against future extreme events.

Part IV explores the consequences of economic activities that can have extreme effects on the environment and other ocean users and also highlights promising new uses of the oceans' natural resources, particularly for the blue economy. Aquaculture is explored as one opportunity, but with the potential to cause negative impacts; and the deep sea is highlighted as an area for development but also an extreme environment about which little is known. Illegal and extreme human behaviours are explored to highlight how people impact on the environment and society as well as enterprises and activities conducted in the ocean.

In Chapter 10, 'Aquaculture', Pomeroy and Sequeira explore one of the world's fastest-growing food production sectors. All over the world, aquaculture is being viewed as an essential part of the solution in addressing food security concerns. It occurs at a broad range of production scales, from small family-owned farms to large commercial operations, and encompasses the culture of a wide variety of marine species including fish, shellfish, sea cucumbers and seaweed. Aquaculture techniques are also being trialled as a way of restoring degraded ecosystems. The authors explore the barriers to further expansion of aquaculture operations: negative impacts due to the intensity of operations, extreme short and long term changes in marine climate and the extreme challenges of offshore operations.

In Chapter 11, Lindley, Techera and Webster explore a range of issues in 'Extreme human behaviours affecting marine resources and industries'. The focus of the chapter is illegal activities which can be damaging to the marine environment and its resources as well as to people and industry. The chapter explores three critical examples: illegal, unreported and unregulated fishing; human trafficking and forced labour onboard vessels; and maritime piracy. These crimes all involve extreme human behaviours and are often linked to broader transnational organised crime and significant maritime security issues. This chapter analyses efforts to address these extreme activities, the underlying motivators and facilitators, and ways in which outcomes might be enhanced in the future.

In the final chapter in Part IV, Collin, Chapuis and Michiels provide a fascinating examination of the 'Impacts of (extreme) depth on life in the deep-sea' (Chapter 12). The deep sea represents the most extreme environment on Earth and remains largely unexplored. The authors skilfully explain what is known about this extreme environment and the uniquely adapted organisms that live there. Although logistically and financially difficult to explore and monitor, the health of the deep sea is paramount to maintaining stable meteorological conditions, preserving this critical marine ecosystem and predicting the effects of future climate variability. The authors emphasise that there is a dire need to understand more about the physical and biological drivers of maintaining the health of the deep sea, particularly given increased interest in achieving blue economy goals associated with biotechnology and seabed mining.

The final part of the book, Part V, synthesises the bodies of knowledge and case studies explored in the chapters in Parts II, II and IV and draws out the lessons learned and the options and opportunities moving forward. In 'Addressing the challenges and harnessing the benefits of marine extremes' (Chapter 13), Techera and Winter underline the value of theoretical and applied information that is rapidly being acquired in relation to marine extremes, but also the vital need for continued research across diverse disciplines and related to various challenges. To this end they propose a research agenda across the different aspects of marine extremes explored in this volume, as well as to look at how new knowledge and enhanced understanding could be utilised by stakeholders to inform good governance of the oceans, human activities and responses to marine extremes.

Conclusion

This volume draws upon principles from the physical and social sciences to assess the impacts of marine extremes for communities, infrastructure and marine ecosystems. The book reviews processes and forecasting methods to predict marine extremes, assesses monitoring options to gauge the effects of marine extremes and evaluates approaches and measures to mitigate the impacts of marine extremes. It highlights threats posed by marine extremes but also emerging opportunities. While addressing exemplary challenges, consequences and solutions, this volume provides a unique insight into marine extremes – from severe environments and events, to acute human behaviours and impacts of intense anthropogenic activities. All of the disciplines represented and explored in this book are essential in enhancing marine scientific knowledge and embedding it in ocean governance regimes, if the future of our oceans and the goods and services they provide are to be secured. This volume builds multi-disciplinary expertise to contribute to this goal. It is a valuable addition to the literature, of importance to scholars, governments, non-governmental organisations, practitioners and commercial operators.

2   The science, social science and governance of marine extremes

Erika J. Techera and Gundula Winter

Introduction

Relatively little is known about the marine environment compared to land areas. Whilst coastal areas and estuaries are well-mapped and marine life more comprehensively recorded, the deep ocean remains largely a mystery. Advances have been made yet the global map of the ocean floor, for example, is less detailed than that of Mars or Venus (Copley, 2014; Witze, 2014). Furthermore, until recently marine life in extreme depths of the ocean, areas remote from the shore, and hostile environments such as the polar regions were largely unexplored. The marine environment is, in many respects, an alien 'extreme' world distinctly separated from more hospitable land areas that humans inhabit. Yet we are inherently linked to the oceans which we have relied on for subsistence and livelihoods throughout history, as well as for culture, recreation and economic development. Therefore, we should have significant concerns about extreme events and activities that impact upon the health of the ocean and its resources, on which we depend.

Since industrialisation, and more rapidly in the last few decades, scientific advances have enabled us to push further into remote marine areas, and to travel to previously unreachable maritime zones. Progress in engineering and technology, for example, enables ships to travel through very remote waters, manned and unmanned submersible vehicles to explore deep ocean trenches and vessels to remotely track and attract schools of fish. Other fishing technology has allowed greater volumes of seafood to be caught from increasingly large boats with vast nets trawling at extreme depths. Simultaneously, the use of aquaculture has been expanded to yield extreme production quantities in the spatially limited coastal zone. Exploration and extraction of fossil fuels can now be achieved far from the shore in isolated places and in extreme conditions.

These human activities take place in a marine environment characterised by extremes including meteorological, environmental, economic and human behavioural extremes. These extremes pose risks to the health and safety of the marine environment and human activities within it. Where a large number of people, communities, economic activities or natural habitats are directly threatened by marine extremes, the risk to the environment and to human wellbeing is greatest. In remote, isolated and extreme environments, the potential long-term consequences for the marine environment and resources are greater as responses to problems tend to be limited and slower. Simultaneously, other scientific research has made it clear that climate change is increasing the probability of natural extreme events, which puts the health of ocean environments at risk. In particular, marine heatwaves, rising sea levels, acidification and extreme weather events appear to be increasing in frequency and intensity and scientists are working rapidly to better predict and model future trends and their impacts. This new knowledge has financial implications as governments and industry seek new areas for economic development and must grapple with this changing ocean environment.

Approximately 5 per cent of global gross domestic product (GDP) is currently related to oceans (Narula, 2017). Around the world we have seen the emergence of blue economy goals incorporating multiple objectives; these include increasing wealth from the oceans across fishing, shipping, energy, biotechnology and tourism sectors. Blue economy goals must also be achieved in sustainable ways to ensure long-term outcomes, but they too are threatened by current and emerging marine extremes. There is little doubt that good oceans governance is essential if the blue economy agenda is to be achieved. Efficient and effective laws and processes administered by responsible and equitable institutions are essential if development is to be balanced with environmental protection. At the same time, regard must also be given to socio-cultural matters including the impacts of a rapidly changing ocean environment on people and communities, and contributions that bottom-up, participatory approaches can make to addressing the challenges. The combined efforts of governments, industries and communities are needed to build resilience to marine extremes for the marine environment and human activities.

This chapter explores the multi-faceted nature of knowledge about and activities in our oceans. The role of science is explored: where scientific knowledge is emerging, or needed, and how it can be applied to address and adapt to extremes in the ocean. Thereafter, economic implications and approaches are examined including those related to the blue economy agenda. Top-down and bottom-up approaches to oceans governance are analysed as well as how scientific, economic and socio-cultural concerns must be balanced to achieve blue economy goals and a sustainable future for our oceans.

Science and marine extremes

The oceans are an environment of extremes: from low temperatures at the poles to high temperatures at the equator and from shallow coastal waters to the deep-sea. Our oceans also experience a myriad of environmental extreme events such as storms, hurricanes and marine heatwaves as well as cumulative stressors such as pollution and acidification. Solid scientific knowledge on these extreme events and cumulative processes is critical to understand their impacts on the natural, socio-cultural and economic environment and to manage those impacts. This effort requires scientists from multiple disciplines to work together to understand the pathways and impacts of marine extremes in order, ultimately, to model and forecast them for management purposes.

The sun fuels all atmospheric processes and controls temperature, atmospheric pressure and wind on Earth. These atmospheric processes drive currents and waves in the ocean, which in turn have a feedback effect on the atmosphere. In fact, deep ocean circulation redistributes heat from the equator to the poles and thus regulates the Earth's climate (Stewart, 2008). Meteorologists study these atmospheric processes to provide long-term climate projections and short-term weather forecasts. Physical oceanographers study the ocean and use this atmospheric data, such as temperature, pressure and wind, to understand and model currents, waves and mixing in the oceans. This knowledge enables them to understand and forecast extreme events such as storms and tsunamis, which unleash extreme forces within the environment. These can destroy not only built infrastructure but also benthic organisms such as corals, seagrass and kelp, which provide a habitat to other marine life. Physical processes in the oceans and atmosphere are also key to many chemical and biological processes. For example, in upwelling regions currents bring nutrient-dense water up to the surface where, in combination with sunlight, the production of chlorophyll is triggered. Overall, the oceans contribute 50 per cent to primary production on Earth (Field *et al.*, 1998) and are thus an important oxygen contributor. Primary production hotspots attract higher trophic species and form the basis for rich marine life as higher trophic species can feed on the primary producers. But the atmospheric and oceanic processes also govern the development of marine heatwaves, which are triggered by climate phenomena such as El Niño where a change in wind patterns over the Southern Pacific leads to warmer than usual water temperatures. Ocean currents then transport this warm water to other ocean areas where it often has a damaging effect on local species and ecosystems (see Chapter 8). Heatwaves are exacerbated by global warming which, alongside other cumulative processes such as pollution and acidification, has a global impact on the marine environment. The First Census of Marine Life was a decade long effort to create an inventory of marine life, in which 2,600 mainly biological scientists participated. This inventory has concluded that many marine environments are already impacted by human-induced and natural marine extremes, primarily in coastal seas (Williams *et al.*, 2011). With respect to these developments, scientific research has framed alarming messages for the public: on the current trajectory, coral reefs will die or severely change by the end of this century (Hughes *et al.*, 2017), plastic particles are already found in the remotest and most extreme marine environments including the deep-sea (Woodall *et al.*, 2014) and the Arctic (Obbard *et al.*, 2014), and biodiversity loss is accelerating (Secretariat of the Convention on Biological Diversity, 2014).

These emerging issues require scientists to work together across many disciplines. The conservation of marine ecosystems requires ecologists and biologists to understand the traits of different species and their interrelations. Oceanographers and climatologists contribute long-term projections of extreme events that will threaten ecosystem functioning. The example of plastic pollution as a cumulative extreme demonstrates the necessity for a multi-disciplinary approach. Oceanographic models can help in understanding transport and accumulation of plastics from predominantly land-based and marine-based sources. Biologists and ecologists investigate the impacts of macro and micro plastics on organisms, food webs and ecosystems. Chemists and toxicologists study the toxic material that adheres to plastic particles. As micro plastics enter the food web and sea food is ultimately consumed by humans this is also an issue requiring attention by medical research. In Indonesia and the US, anthropogenic debris is already found in more than half of all fish species (Rochman *et al.*, 2015). Extreme events exacerbate the issue of plastic pollution, such as during the 2011 tsunami off the coast of Japan when a large amount of debris was washed into the Pacific Ocean (Lebreton and Borrero, 2013). Engineers can contribute solutions that diminish or at least reduce plastic pollution at its source, for example, through recycling, better waste water treatment and the collection of particles from the oceans and rivers. As this is a human-induced issue, social scientists are crucial to help change behavioural patterns and to implement better management strategies.

Many economic activities rely on the ecosystem services that the oceans provide. To utilise these services in a sustainable manner, scientific knowledge about natural processes and feedbacks is critical. Currently, many fisheries are fully exploited or over-exploited (FAO, 2016) due to the extreme scale and practices of the fishing industry. Biologists and ecologists can work together to define sustainable fishing quotas and to design marine protected areas or closure times for fisheries. While such restrictions lower profits for fisheries in the short-term, fish stocks improve so that fishers benefit in the long-term (Cinner *et al.*, 2006). Another win–win situation for marine conservation and economic development is apparent in eco-tourism. Eco-tourism provides a mechanism to earn profits from natural resources repeatedly as opposed to extractive methods such as fishing, which only yield a one-off profit. This raises awareness of marine species and their status and can build stewardship ethics and values associated with marine species and environments. It can also help finance marine conservation, which increases the resilience of ecosystems to extreme events (see Chapter 5). Oceanographers, along with meteorologists, are continually improving forecasting models to provide better services for shipping, recreational activities and search and rescue, as well as offshore operations so human activities can take place safely without being impacted by marine extreme events. Engineers are required to design offshore and coastal infrastructure as well as vessels to withstand extreme events while still being cost-efficient during construction and maintenance. This is important to minimise economic loss and harm to human and environmental health. The explosion on the BP 'Deepwater Horizon' and subsequent oil spill demonstrated the enormous risks to the marine environment that originate from human offshore activities during extreme events. Shipping bears the risk of introducing invasive species (through bio-fouling and in ballast water) and ecologists are required to understand the impact of invasive species on marine environments; these threaten 36 per cent of International Union for Conservation of Nature (IUCN) Red List species (Kappel, 2005). This knowledge can then be implemented in order to achieve better management of the issue. The impacts of invasive species can be irreversible and costly to manage: for example, controlling the damaging effects of the Zebra mussel invasion in the US, which cost US$1 billion (Pimentel *et al.*, 2005).

Climate change is a cross-cutting issue in respect of all marine extremes. In the recent past, climate change has already caused sea levels to rise and the oceans to warm and to become increasingly acidic. Scientists are working hard to understand future developments and to identify irreversible tipping points in the climate and how this will affect the oceans, marine ecosystems and the humans that depend on them. Sea level rises and ocean warming will accelerate (IPCC, 2014) shifting the baselines and thus exacerbating extreme events such as storm surges and marine heatwaves. Research also suggests that the frequency and intensity of weather phenomena such as El Niño are likely to increase, which could cause marine heatwaves with disastrous impacts on the marine environment, coastal communities and economies. More frequent marine heatwaves leave little recovery time for coral reefs, kelp forests and other important ecosystems. These ecosystems provide nurseries for fish, which are crucial to industrial, recreational and artisanal fisheries, as well as other ecosystem services. A loss of coral reefs would deprive approximately 500 million people worldwide of their livelihoods and coastal protection (Wilkinson, 2004). Singular events such as tropical cyclones can render the efforts of years of marine conservation useless within a matter of hours. And an increase in storm frequency and intensity due to climate change will leave less time for ecosystems and communities to recover from these events. Biologists and ecologists now aim to find out to what extent species and ecosystems can resist and adapt to changing extreme events due to climate change. In the meantime, science and engineering are working hard on solutions to mitigate climate change, including carbon capture by algae and marine renewable energy including offshore wind and tidal and wave energy.

Collaboration between all scientific disciplines is critical to understanding the feedback between physical, chemical, ecological and biological processes and to making predictions of future impacts. But as the examples above highlight the sciences are not the only relevant body of knowledge in order to safeguard ocean health and sustainable development in the light of changing extreme events. Social sciences, particularly human geography, economics, law, political science and sociology, are equally important in implementing possible solutions and to bring about change.

Economic implications of marine extremes

Our oceans have been utilised for economic gain for millennia. The very earliest humans ate food from the oceans and later community-based activities were undertaken and marine resources bartered or traded. We continue to make economic products from raw marine materials. Oceans have been used for exploration and navigation to find new lands and for trading and shipping routes. Oceans are in essence economic commercial highways with 80 to 90 per cent of the world's trade transported by sea (Narula, 2017; World Bank and United Nations Department of Economic and Social Affairs, 2017). Industrialisation led to the expansion of ocean-based industries, such as commercial fisheries and later energy from the oceans, at ever increasing scales. The Food and Agriculture Organization (FAO) estimates that there are 4.6 million fishing vessels in the world and that fish provide 17 per cent of the global population's annual consumption of animal protein (FAO, 2016). More recently, non-consumptive exploitation of the oceans has seen a rise in marine-based tourism and recreational activities involving over 121 million people globally (Cisneros–Montemayor and Sumaila, 2010).

Extractive industries, such as fishing and oil and gas production, involve the commodification of raw materials. When these resources are removed from the ocean the natural environment can be diminished because the structural building blocks of ecosystems are removed (Farley, 2012). Furthermore, the process of production itself can cause environmental damage as waste is produced. In a variety of ways, marine extremes affect the ecosystem goods and services that the oceans provide and therefore economic aspects of ocean-based activities. Illegal extreme behaviours such as piracy and robbery at sea, illegal fishing and transnational crime all impact directly on financial bottom lines but also, in many respects, on the marine environment itself (these aspects are considered further in Chapter 11). Extreme weather events and the effects of climate change impact upon the economic gains to be made from such oceanic activities, and the enterprises themselves can have extreme and costly impacts in the form of pollution and environmental degradation, as explored in Chapter 7. The estimated economic impact of the 2004 Asian Tsunami in India and Indonesia, for example, was US$500 million (FAO, 2016).

Economic theories and approaches to valuing ecosystem services have been studied (Silvis and van der Heide, 2013). Nevertheless, not all of the ecosystem services provided by the oceans and marine resources have been fully incorporated into decision-making – in particular the productive and sink functions. This failure to take into account the triple bottom line – economic, environmental and social costs – is at least partly responsible for unsustainable activities extending beyond the carrying capacity of natural environments, species and ecosystems. While economic development based on oceans and marine resources has always occurred, and potentially could be conducted within natural limits, it has become increasingly evident that the scale of activities, and the lack of true accounting for environmental impacts, can have unsustainable consequences. For example, advances in technology have allowed fishing to take place in extreme environments with little account taken of the negative biological and ecological consequences, as can be seen with bottom trawling. Profitable trading routes have only been made possible because of the oceans but quicker and riskier routes have real and potential environmental impacts. Oil and gas exploration, development and transport occur in harsh oceanic environments from the North Sea to the Southern Ocean, with increased risks of incidental damage and accidents with catastrophic consequences. Sea bed mining is one example of an emerging commercial activity that is conducted at extreme depths with little known impacts making the true cost difficult to evaluate. Oceans are not endless suppliers of resources nor endless waste sinks – there are thresholds and limits – and therefore the cost of these services ought to be internalised in business models to ensure sustainable use of natural capital.

Our reliance upon oceans can be seen at all scales from local to global. Globally, it is estimated that ocean, coastal marine and estuarine areas contribute approximately US$50 trillion per year in ecosystem services (Costanza, 2014). In some countries, significant proportions of GDP are derived from marine resources and ocean environments. In the US, for example, one-third of GDP originates in coastal areas, in China ocean commerce is approximately 10 per cent of GDP (Zhao *et al.*, 2014) and in Indonesia oceans provide around 8 per cent of GDP. As land-based resources become depleted attention has again turned to the oceans, leading to the emergence of the concept of the blue economy.

The 'blue economy' was a term first used by Gunter Pauli and marked a paradigm shift in thinking. In essence, the concept of the blue economy involves a way of addressing multiple environmental issues, based upon the needs of all living things, utilising innovative business models that take advantage of untapped potential and result in zero waste – it is not only focussed on oceans (Pauli, 2010). The term 'blue economy' has also been utilised by governments and inter-governmental organisations, but with various interpretations (Voyer *et al.*, 2018). For example, the Indian Ocean Rim Association (IORA) adopted the *Jakarta Declaration on Blue Economy* with a heavy emphasis on promoting economic growth alongside alleviating poverty, securing food and ensuring sustainable use of the oceans. Increasingly, national growth agendas have emerged with a strong focus on the blue economy, as well as 'blue growth' and 'ocean economy', which are equally poorly defined (Eikeset *et al.*, 2018). Although policies vary, wealth from the oceans is a common feature (see, for example, Seychelles) (Commonwealth Secretariat, 2018). The concept of the blue economy has also attracted attention from other bodies including the World Bank and Organisation for Economic Co-operation and Development (OECD) (OECD, 2016; Patil *et al.*, 2016). The risk for the blue economy is that wealth from the oceans, including marine resource-based products, will be prioritised over ecosystem needs leading to resource utilisation beyond the carrying capacity of the resources. In addition, cross-cutting issues such as climate change that result from unsustainable anthropogenic activities will undermine potential gains. At the global level, the United Nations has sought to re-focus attention on marine health through the Sustainable Development Goals (SDGs), discussed in more detail below. Economic approaches are important in determining the financial cost of climate change mitigation, for example, as well as damage assessment and compensation. However, as will be explored below, economic valuations alone will not determine the overall costs of the impacts of marine extremes or the extreme impacts of anthropogenic activities. Economic aspects are not only relevant to private industry and government sectors but are important to people and communities. Values do not only include economic ones and, as will be seen in the next section, non-market and non-economic values are key considerations in working within, adapting to and addressing marine extremes.

Socio-cultural concerns and contributions

As noted above, humans are inextricably connected with the ocean from evolutionary, cultural and societal perspectives. Life originated in the ocean and the earliest human communities were found around coastal areas. Many coastal communities in developing countries and small island states continue to have lifestyles that are at least in part traditional; fish consumption is largely based on locally caught seafood which provides 50 per cent or more of total animal protein (FAO, 2016). Small scale fisheries provide livelihoods for 90 per cent of people engaged in capture fisheries globally, without which many would have few other employment options (FAO, 2016). In more industrialised, urbanised environments, people still rely on the oceans for protein in the form of seafood which may be locally harvested or imported from large distances. Fishing, ports, shipping and the oil and gas industry all contribute to local culture as well as employment. Dependence upon the ocean is not limited to food, livelihoods and commodities. In almost all coastal environments people are drawn to the ocean for cultural and recreational purposes. In some cases, this has resulted in the development of marine-based tourism enterprises and recreational activities at all scales and involving extreme sports, ecotourism and research tourism (Cisneros-Montemayor and Sumaila, 2010). Furthermore, many nations and peoples have cultures, traditions and legends associated with the oceans, and specific beliefs, fears and reverence for both species and marine phenomena (see World Ocean Observatory – http:// worldoceanobservatory.org).

The outline above demonstrates the range of different connections people and communities have with the oceans and marine resources. In order to understand these linkages and interdependencies, social science research is essential. Fields such as history and archaeology can tell us about past uses of the ocean and how societies have responded to changing marine environments. Human geography examines how humans affect and are affected by the environment. Anthropology, sociology and fields such as cultural studies explore the functioning of human societies. Philosophy provides an understanding of values and ethical attitudes including those related to nature. Psychology explores how humans behave and their mental processes, which also contributes to an understanding of attitudes towards the environment and ways in which to address extreme human behaviours. Law, policy and political science research is critical in managing people and controlling human activities. Together these fields cover a spectrum of issues directly relevant to human–nature interactions.

It is clear that despite heavy dependence on oceans and marine resources, people and society have had negative impacts on the environment. Extreme effects can be seen in several ways. First, extreme human behaviours (considered further in Chapter 11) in the form of illegal activity can damage the marine environment and undermine economic activities. Piracy, for example, has had financial implications for shipping, and illegal fishing and dumping of waste has damaged the environment.

A second aspect of negative human impacts is marine pollution. It was once believed that the oceans were so vast that we could not have any meaningful impact upon them. It has been clear for some time that this is no longer true. Where populations were small and activities artisanal, impacts were minimal. Increasing industrialisation and growth has resulted in extreme impacts. In some cases, these impacts are acute – such as catastrophic damage caused by oil spills from tankers and rigs. In other cases, the damage is more cumulative. Climate change is one such example as CO_2 levels from multiple sources, countries, communities and individuals combine to have a devastating effect. One of the most recent and pressing challenges is marine debris in the form of plastics. Only in the last few years has the extent of the problem begun to emerge and the global nature of the problem highlighted. In order to overcome such challenges, social science research is needed to identify ways to change our patterns of behaviour. There is no doubt science, engineering and technology will be important to understand the issues and develop potential solutions, but cultural studies, applied ethics, psychology and cognitive behavioural studies will also be critical to facilitating change.

Social licence to operate is also an important concept that has only recently received research attention. Social licence to operate involves businesses conducting their operations within acceptable societal limits and adjusting their practices to avoid community conflict (Gunningham *et al.*, 2004). Traditionally, social licence to operate has referred to expectations imposed on commercial activities, but it has also started to influence government decisions about the use of resources (Cullen-Knox *et al.*, 2017). Particularly where there is the potential for extreme changes to or impacts upon the environment, community participation in decision-making becomes critical. Confidence in regulators can also be lost where governance is not seen to be effective. Public participation is considered further in the section on governance below.

Given the socio-cultural importance of coastal and marine areas and resources for people and communities, social science research has a significant role to play in avoiding extreme human behavioural impacts on the ocean and responding to the effects of extreme events, including those related to climate change, on human society. Social science research is also valuable in responding to climate change. It is clear that, in designing mitigation and adaptation responses to climate change, socio-cultural concerns and impacts must be taken into account in addition to the economic effects. Such effects of climate change will be many and varied: loss of underwater cultural landscapes and archaeology, sacred sites, recreationally important coastal areas, traditional and community-based fishing grounds as species' ranges change. Extreme weather events caused by climate change can have acute negative effects on communities and culture, and slower changes such as acidification and sea level rise will be problematic in the longer term.

Not only must these impacts be taken into account, but socio-cultural input, knowledge and practices have a role to play in working in the context of, or addressing, extremes. Coastal communities are the most 'local' to ocean areas and therefore they hold particular knowledge and information that may be valuable in designing future policies and actions to address climate change and other impacts. For example, in many island states people are so heavily dependent upon marine areas for food, livelihoods and resources that they have a strong stewardship ethic. For many indigenous communities the land–water interface is continuous and attitudes and approaches to the ocean are very different from western constructs. This has translated into different attitudes and approaches to the ocean environment. Addressing marine extremes will require input from these communities as well as governments and industry (as explored in Chapters 4 and 9).

These interests must, however, be valued in ways that enable them to be incorporated into decision-making frameworks; qualitative approaches to analysing socio-cultural values again rely heavily on participatory processes. Economic and non-economic valuation methods can be utilised, most importantly qualitative methods. Such qualitative approaches sometimes stand alone, but in other cases must be utilised in conjunction with quantitative methodologies to present a full cost–benefit analysis for decision-makers. Assessing values is particularly important in the context of marine extremes as it captures the willingness to pay for certain responses – to climate change, for example. Detailed analysis of valuation methods is explored in Chapter 4.

As explored below, the importance of incorporating socio-cultural concerns has been recognised by international, regional and national institutions and incorporated into environmental law, policy and governance. In addition, the principles of good governance require the participation of all stakeholders including local communities, indigenous and traditional peoples and those parts of civil society that do not have economic interests.

Good oceans governance

As explored above, scientific, economic and socio-cultural input is needed in order to achieve the spectrum of blue economy goals. Nevertheless, decision-making in relation to human activities involving oceans and marine resources, and responses to extreme marine environments and events, falls largely to states or governments having the authority to do so. Good governance is therefore critical. Governance has no globally-endorsed definition but has, for example, been explained as 'steering human behaviour through combinations of state, market and civil society approaches in order to achieve strategic objectives' (Jones, 2014, p. 63). For the purposes of this chapter, it is considered to involve the framework of rules, processes and actors involved in decision-making. This top-down approach is appropriate where governments are the key actors responsible for approving marine activities and responding to marine extremes. Whilst bottom-up approaches are relevant, and community-based values and industry sector approaches important, inclusivity and equity can be built into top-down governance regimes. Good governance is best defined through core principles: efficiency, effectiveness, inclusivity, participatory, equity, accountability, responsiveness and transparency. In addition to these principles, adherence to the 'rule of law' is included, which ensures that the same rules apply to all stakeholders within a jurisdiction – governments, industry and individuals – and is a mechanism to avoid corruption. Good governance is essential to ensure ocean health and the achievement of blue economy goals because its application involves all stakeholders as well as appropriate outcomes that respond to the issues being faced.

Good governance largely relates to the processes of governing; for example, who is involved, how decisions are made and what is communicated. Other concepts and principles provide further substantive foundations for law-making and decision-making. A critical concept in this regard is sustainable development: development that meets the needs of the present without compromising the ability of future generations to meet their needs (World Commission on Environment and Development, 1987). Sustainable development itself is comprised of a number of principles including the integration of social, economic and environmental concerns, the precautionary principle, the polluter pays principle, conservation of biological diversity, inter- and intra-generational equity and internalisation of environmental costs. It can quickly be seen that there is considerable overlap between these principles and those of good governance. The original elements of the concept of blue economy are also well-aligned with sustainable development. Collectively,

•   human activities in the ocean and responses to marine extremes must be governed effectively and efficiently founded on science-based information and socio-cultural awareness;

- decision-making must be transparent and integrated to balance sustainable use of the environment and resources with environmental protection and socio-cultural concerns;
- all stakeholders must be engaged inclusively, equitably and in ways that encourage meaningful participation; and
- decision-makers must be accountable for their actions.

Just as scientific understanding of our oceans has lagged behind research on land, so too oceans governance emerged later (Woolley, 2014). Together, good governance and sustainable development provide a strong foundation for sustainable utilisation of marine areas and resources, and the achievement of blue economy goals. This has been recognised by the international community and national governments in laws and policies focussed on oceans governance and the management of marine extremes. At the inter-governmental level, the global community has adopted an oceans-focussed Sustainable Development Goal (SDG). SDG 14 – 'Life Below Water' – refers specifically to several marine extremes including climate change, overfishing and marine pollution and the need to increase scientific knowledge and economic benefits while sustainably managing resources, minimising impacts and building resilience (see Sustainable Development Goal 14 – https://sustainable development.un.org/sdg14).

Oceans governance at the global level includes international treaties and conventions, adopted and implemented under the auspices of inter-governmental organisations. Relevant sub-fields include law of the sea, fisheries law, pollution law, conservation law and disaster law. Each area has its own inter-governmental organisation that takes the lead and international legal frameworks to address activities and prevent impacts. The most relevant inter-governmental bodies include the FAO with responsibility for fisheries; the International Maritime Organization (IMO), which sets standards for shipping and some oil and gas regulation; and the United Nations Environment Programme (UNEP), which has taken a lead role in developing pollution treaties and regional seas agreements.

The key international law relating to marine areas and resources is the *UN Convention on the Law of the Sea* which establishes maritime zones (territorial sea, exclusive economic zone and high seas) as well as rights to fish in certain waters and general obligations to 'protect and preserve' the marine environment, designed to prevent negative impacts from over-use of marine resources and extreme human behaviours. Fisheries laws set allowable catches, determine species that may be caught and their appropriate sizes, restrict the use of certain equipment and set seasons and areas where fishing can take place; in essence creating a regime for legitimate fishers and establishing the parameters of illegal fishing. This is done through international agreements (*UN Fish Stocks Agreement, FAO Compliance Agreement, Port State Measures Agreement*) as well as regional fishery management organisations (RFMOs) such as the Indian Ocean Tuna Commission (IOTC), North East Atlantic

Fisheries Commission (NEAFC) and Western Central Pacific Fisheries Commission (WCPFC). Conservation treaties relevant to the ocean environment include the *Convention on Biological Diversity, Convention on Migratory Species* and the *Convention on International Trade in Endangered Species* which provide further protection from deliberately damaging extreme human behaviours and anthropogenic activities, as well as for species with poor conservation status due to other marine extremes. International law regimes designed to prevent extreme impacts such as marine pollution (*MARPOL 73/78* and the *London Convention*) prohibit the dumping of waste and require operators to put in place measures to prevent incidental pollution. Extreme marine environments such as Antarctica and the sea floor in areas beyond national jurisdiction (i.e. beneath the high seas) are the common heritage of humankind, unable to be owned or utilised by one state alone. The International Seabed Authority (ISA) regulates mining on the sea floor, including assessing the environmental impacts and approving permits.

In terms of responding to extreme marine events, disaster law is a relatively new sub-field that focuses largely on risk reduction, relief and responses to catastrophes such as landslides, tsunamis and earthquakes, and brings humanitarian and human rights law together with some migration and environmental law (see International Federation of Red Cross and Red Crescent Societies – www.ifrc.org/en/what-we-do/disaster-law/). In this sense, much of the work in the area of disaster law addresses the aftermath of events, including marine extremes. More recently, the United Nations Office for Disaster Risk Reduction (UNISDR) has served as a focal point for coordination of disaster risk reduction activities, particularly those associated with climate change (see www.unisdr.org/). In 2015, the United Nations General Assembly adopted the *Sendai Framework for Disaster Risk Reduction 2015–2030* to build understanding of risk, strengthen governance, encourage investment in disaster risk reduction and improve preparedness in anticipation of extreme events. The focus is strongly on reducing impacts on and mortality of people and protecting against economic loss and damage to infrastructure, as well as improving cooperation and warning systems.

Equally relevant is climate change law and policy which seeks to address underlying causes, as well as facilitating adaptation strategies. The *UN Framework Convention on Climate Change, Kyoto Protocol* and *Paris Agreement*, together with the work of the Intergovernmental Panel on Climate Change (IPCC), create obligations to reduce greenhouse gas emissions and mitigate and adapt to the effects of climate change. Under the *Paris Agreement*, for example, countries must commit to a range of 'nationally determined contributions' through mitigation and adaptation to climate change (see United Nations, Climate Change – https://unfccc.int/process-and-meetings/the-paris-agreement/nationally-determined-contributions-ndcs). Mitigation can include emissions reductions through renewable energy initiatives to reduce reliance on fossil fuels, and/or 'blue carbon' schemes that value and protect coastal marine ecosystems for the sink functions they provide (see the Blue

Carbon Initiative – http://thebluecarboninitiative.org). Adaptation commitments include protection of coastal infrastructure, food and water security projects and disaster preparedness. A number of climate change governance challenges remain: for example, compensation regimes for damage and migration scenarios.

International law sets standards, creates frameworks for global cooperation on common issues and is a mechanism to share knowledge and expertise. It therefore plays an essential role in addressing global issues such as climate change. In most cases international law does not operate automatically. Individual states sign treaties and must then pass domestic legislation and establish national authorities to implement their obligations. So, while international law can establish harmonised rules, set standards and facilitate the sharing of knowledge and expertise, most regulatory agencies and measures are at the national level. It is clear that different scales of governance are relevant for different issues. For example, regional law and policy has a role to play where specific issues might affect a geographic area encompassing the territory of several states. This is particularly the case in oceans governance where, for example, RFMOs adopt measures for specific fisheries across an ocean within exclusive economic zones (EEZs) and on the high seas. Similar approaches have been taken in relation to regional pollution issues through the regional seas conventions and bodies. Tailoring international obligations to local contexts involves the participation of all stakeholders and bodies of knowledge, regular monitoring and surveillance.

Law is critically important in addressing the impacts of human activities on the ocean. Law can provide a 'back stop' to disincentivise and punish deliberately damaging behaviour. Yet, despite all of the above principles and concepts, institutions and legal frameworks, ocean indicators have continued to decline (Halpern *et al.*, 2009; Halpern *et al.*, 2012; Ocean Health Index). It is clear that there have been considerable successes in discrete areas: oil spills, for example, are at record lows (Jernelöv, 2010); but multi-faceted and complex ocean issues are not ideally suited to the compartmentalised governance approach taken. The global politico-legal system focuses on sovereignty and rights based on territorially delimited areas. Jurisdictional boundaries in the ocean, as well as the land–sea divide, are mismatched with ecosystem scales, leading to calls for a stronger focus on ecological governance (Woolley, 2014). Furthermore, issues that cut across a number of areas, such as food, maritime and human security, require input and action by a number of institutions and actors, as well as the incorporation of different bodies of knowledge and best practice concepts.

Climate change is one such cross-cutting issue that not only has contributions from, but impacts upon, shipping, fishing, tourism, the energy sector as well as the health of the environment in general, and thus requires input from a range of disciplines. The embedding of scientific concepts in oceans governance is patchy with limited implementation of, for example, ecosystem-based management; yet adaptive governance is as important as adaptive management (Woolley, 2014). Scientific monitoring and modelling is not mandated in most legal regimes but is essential in predicting the future impacts of marine extremes (see Chapter 6). Scientific knowledge of nature-based as well as man-made solutions to coastal impacts, and community support for the different options, must be understood (see Chapters 4 and 5). Similarly, extreme human behavioural issues require legal responses but also exploration of the underlying motivators and facilitators of crime (see Chapter 11). Such multi-disciplinarity is theoretically possible but practically hard to achieve in governance regimes.

Conclusion

The science is clear: we are living beyond our carrying capacity in terms of depleting marine resources and relying too heavily on ocean sinks, and our activities are having extreme impacts both directly and through exacerbation of natural marine extremes via climate change. As populations continue to grow our dependence upon the oceans is unlikely to subside and therefore ways must be found to better balance environmental, socio-cultural and economic concerns. Simultaneously, increasing populations are at growing risk due to extreme events including storms, tsunamis and marine heatwaves, which affect human lives and livelihoods. This chapter has shown that scientific concepts such as ecosystem-based management, can be aligned with approaches to good governance, environmental law and sustainable development. They each contribute common elements including responsiveness, participation and integration. Global consensus has been built around SDG14 which can be interlinked with blue economy goals that themselves have broad support.

The framework is therefore available to address our activities that may have extreme impacts in the ocean as well as the extreme impacts of the oceans upon us. All that remains is the translation of these concepts and approaches into action. Just as scientific data is needed to better model and predict the future of marine extremes, knowledge and information is needed from other areas to advance best practice governance. The chapters that follow explore specific aspects of, and case studies drawn from, science and social science research to demonstrate the way forward to ensure ocean health, human safety and wealth from our oceans for generations to come.

References

Cinner, J., Marnane, M.J., McClanahan, T.R. and Almany, G.R. (2006). 'Periodic closures as adaptive coral reef management in the Indo-Pacific'. *Ecology and Society*, vol 11, no 1, doi:10.5751/ES-01618-110131.

Cisneros-Montemayor, A.M. and Sumaila, U.R. (2010). 'A global estimate of benefits from ecosystem-based marine recreation: Potential impacts and implications for management'. *Journal of Bioeconomics*, vol 12, no 3, pp. 245–268.

Commonwealth Secretariat (2018). 'Seychelles' blue economy – Strategic policy framework and roadmap: charting the Future (2018–2030)'. Retrieved from https://seymsp.com/wp-content/uploads/2018/05/CommonwealthSecretariat-12pp-RoadMap-Brochure.pdf.

Copley, J. (2014). 'Just how little do we know about the ocean floor?'. *The Conversation*, 9 October 2014. Retrieved from https://theconversation.com/just-how-little-do-we-know-about-the-ocean-floor-32751.

Costanza, R. (2014). 'Changes in the global value of ecosystem services'. *Global Environmental Change*, vol 26, pp. 152–158, doi:10.1016/j.gloenvcha.2014.04.002.

Cullen-Knox, C., Haward, M., Jabour, J., Ogier, E. and Tracey, S.R. (2017). 'The social licence to operate and its role in marine governance: Insights from Australia'. *Marine Policy*, vol 79, pp. 70–77, doi:10.1016/j.marpol.2017.02.013.

Eikeset, A.M., Mazzarell, A.B., Davídsdóttir, B., Klinger, D.H., Levin, S.A., Rovenskaya, E. and Stenseth, N.C. (2018). 'What is blue growth? The semantics of "sustainable development" of marine environments'. *Marine Policy*, vol 87, pp. 177–179, doi:10.1016/j.marpol.2017.10.019.

FAO (2016). 'The state of world fisheries and aquaculture 2016. Contributing to food security and nutrition for all'. FAO, Rome.

Farley, J. (2012). Ecosystem services: The economics debate. *Ecosystem Services*, vol 1, no 1, pp. 40–49, doi:10.1016/j.ecoser.2012.07.002.

Field, C.M., Behrenfeld, J., Randerson, J.T. and Falkowski, P. (1998). 'Primary production of the biosphere: Integrating terrestrial and oceanic components'. *Science*, vol 281, no 5374, pp. 237–240, doi:10.1126/science.281.5374.237.

Gunningham, N., Kagan, R.A. and Thornton, D. (2004). 'Social license and environmental protection: Why businesses go beyond compliance'. *Law & Social Inquiry*, vol 29, no 2, pp. 307–341, doi:10.1111/j.1747-4469.2004.tb00338.x.

Halpern, B.S., Walbridge, S., Selkoe, K.A., Kappel, C.V., Micheli, F., D'Agrosa, C., Bruno, J.F., Casey, K.S., Ebert, C., Fox, H.E., Fujita, R., Heinemann, D., Lenihan, H.S., Madin, E.M.P., Perry, M.T., Selig, E.R., Spalding, M., Steneck, R. and Watson, R. (2009). 'A global map of human impact on marine ecosystems'. *Science*, vol 319, no 5865, pp. 948–952, doi:10.1126/science.1149345.

Halpern, B.S., Longo, C., Hardy, D., McLeod, K.L., Samhouri, J.F., Katona, S.K., Kleisner, K., Lester, S.E., O'Leary, J., Ranelletti, M., Rosenberg, A.A., Scarborough, C., Selig, E.R., Best, B.D., Brumbaugh, D.R., Chapin, F.S., Crowder, L.B., Daly, K.L., Doney, S.C., Elfes, C., Fogarty, M.J., Gaines, S.D., Jacobsen, K.I., Karrer, L.B., Leslie, H.M., Neeley, E., Pauly, D., Polasky, S., Ris, B., St Martin, K., Stone, G.S., Sumaila, U.R. and Zeller, D (2012). 'An index to assess the health and benefits of the global ocean'. *Nature*, vol 488, pp. 615–620.

Hughes, T.P., Barnes, M.L., Bellwood, D.R., Cinner, J.E., Cumming, G.S., Jackson, J.B., Kleypas, J., Van De Leemput, I.A., Lough, J.M., Morrison, T.H. and Palumbi, S.R. (2017). 'Coral reefs in the Anthropocene'. *Nature*, vol 546, pp. 82–90, doi:10.1038/nature22901.

IPCC (2014). 'Climate change 2014: Synthesis report. Contribution of working groups I, II and III to the fifth assessment report of the Intergovernmental Panel on Climate Change' [Core writing team, R.K. Pachauri and L.A. Meyer (eds.)]. IPCC, Geneva.

Jernelöv, A. (2010). 'The threats from oil spills: Now, then, and in the future'. *AMBIO*, vol 39, pp. 353–366, doi:10.1007/s13280-010-0085-5.

Jones, P.J.S. (2014). *Governing Marine Protected Areas*. Oxford: Routledge.

Kappel, C.V. (2005). 'Losing pieces of the puzzle: Threats to marine, estuarine, and diadromous species'. *Frontiers in Ecology and the Environment*, vol 3, no 5, pp. 275–282, doi:10.1890/1540-9295(2005)003[0275:lpotpt]2.0.co;2.

Lebreton, L.C.M. and Borrero, J.C. (2013). 'Modeling the transport and accumulation floating debris generated by the 11 March 2011 Tohoku tsunami'. *Marine Pollution Bulletin*, vol 66, no 1, pp. 53–58, doi:10.1016/j.marpolbul.2012.11.013.

Narula, K. (2017). 'Blue Economy and Sustainable Development Goals: Aligned and Mutually Reinforcing Concepts', in V. Sakhuja and K. Narula (eds.) *The Blue Economy: Concept, Constituents and Development.* New Delhi: Pentagon Press.

Obbard, R.W., Sadri, S., Wong, Y.Q., Khitun, A.A., Baker, I. and Thompson, R.C. (2014). 'Global warming releases microplastic legacy frozen in Arctic Sea ice'. *Earth's Future*, vol 2, no 6, pp. 315–320, doi:10.1002/2014ef000240.

Ocean Health Index. www.oceanhealthindex.org/.

OECD. (2016). *The Ocean Economy in 2030.* Paris: OECD Publishing.

Patil, P.G., Virdin, J., Diez, S.M., Roberts, J. and Singh, A. (2016). 'Toward a blue economy: A promise for sustainable growth in the Caribbean: An overview'. The World Bank, Washington, DC.

Pauli, G. (2010). *The Blue Economy.* Taos, NM: Paradigm Publications.

Pimentel, D., Zuniga, R. and Morrison, D. (2005). 'Update on the environmental and economic costs associated with alien-invasive species in the United States'. *Ecological Economics*, vol 52, no 3, pp. 273–288, doi:10.1016/j.ecolecon.2004. 10.002.

Rochman, C.M., Tahir, A., Williams, S.L., Baxa, D.V., Lam, R., Miller, J.T., Teh, F.-C., Werorilangi, S. and Teh, S.J. (2015). 'Anthropogenic debris in seafood: Plastic debris and fibers from textiles in fish and bivalves sold for human consumption'. *Scientific Reports*, vol 5, 14340, doi:10.1038/srep14340.

Secretariat of the Convention on Biological Diversity (2014). 'Global biodiversity outlook 4'. CBD, Montréal.

Silvis, H.J. and van der Heide, C.M. (2013). 'Economic viewpoints on ecosystem services'. Statutory Research Tasks Unit for Nature and the Environment (WOT Natuur & Milieu), Wageningen.

Stewart, R.H. (2008). *Introduction to Physical Oceanography.* Department of Oceanography, Texas A&M University. Retrieved from https://ocean.tamu.edu/academics/resources/ocean-world/resources/Stewart_PObook.pdf.

Voyer, M., Qirk, G., McIlgorm, A. and Azmi, K. (2018). 'Shades of blue: What do competing interpretations of the blue economy mean for oceans governance?'. *Journal of Environmental Policy & Planning*, vol 20, no 5, pp. 595–616, doi:10.1080/1 523908X.2018.1473153.

Wilkinson, C. (2004). 'Status of coral reefs of the world: 2004'. Australian Institute of Marine Science, Townsville, Queensland.

Williams, M., Mannix, H., Yarincik, K., Miloslavich, P. and Crist, D.T. (2011). 'Scientific results to support the sustainable use and conservation of marine life: A summary of the Census of Marine Life for decision makers'. Census of Marine Life International Secretariat, Washington, DC.

Witze, A. (2014). 'Gravity map uncovers sea-floor surprises'. *Nature News*, 2 October 2014. Retrieved from www.nature.com/news/gravity-map-uncovers-sea-floor-surprises-1.16048.

Woodall, L.C., Sanchez-Vidal, A., Canals, M., Paterson, G.L.J., Coppock, R., Sleight, V., Calafat, A., Rogers, A.D., Narayanaswamy, B.E. and Thompson, R.C.

(2014). 'The deep sea is a major sink for microplastic debris'. *Royal Society Open Science*, vol 1, no 4, doi:10.1098/rsos.140317.

Woolley, O. (2014). *Ecological Governance*. Cambridge: Cambridge University Press.

World Bank and United Nations Department of Economic and Social Affairs (2017). 'The potential of the blue economy: Increasing long-term benefits of the sustainable use of marine resources for small island developing states and coastal least developed countries'. World Bank, Washington DC.

World Commission on Environment and Development (1987). *Our common future* (Annex to UN Doc A/42/427). Oxford: Oxford University Press.

Zhao, R., Hynes, S. and He, G.S. (2014). 'Defining and quantifying China's ocean economy'. *Marine Policy*, vol 43, pp. 164–173, doi:10.1016/j.marpol.2013.05.008.

Part II
Safe oceans

3 Coastal processes, extreme events and forecasting

Gundula Winter, Yasha Hetzel,
Peisheng Huang, Matthew R. Hipsey,
Ryan P. Mulligan and Jeff Hansen

Introduction

The coastal zone is at the interface between the land and the ocean and is affected by marine, terrestrial and atmospheric processes. Many coastal environments provide unique natural habitats, but also ideal conditions for human settlement and commercial operations. As many economic activities including resource extraction and trade take place in the coastal zone it attracts a population three times denser than the global average (Small and Nicholls, 2003). Coasts can be composed of a combination of characteristics from rocky to sandy and vegetated environments. Sandy shores are a prevalent coastal feature throughout the world and are of importance to the tourism industry in many countries. Sandy shores are also highly dynamic and are constantly being reshaped by waves and currents. River deltas, for example, deliver large amounts of sediment to the coast and advance it seaward. Deltas are also the gateway to the coastal hinterland and are thus important intersections on trade routes, for example, Rotterdam in the Netherlands and Shanghai in China. Rocky coasts have more stable shorelines and provide substrate that organisms can adhere to. They can thus support diverse subtidal and intertidal habitats. In shallow coastal waters, vegetation and other organisms grow, such as seagrass, saltmarsh, mangroves and coral reefs that can protect the coast from negative impacts (see Chapter 5) and are habitat builders for a range of ecosystems. These ecosystems are important for the livelihoods of coastal communities as they form the basis for fishing, tourism and recreation. Estuaries are highly productive aquatic ecosystems and their typically brackish water (mix of salt and freshwater) makes many of these ecosystems unique. The sheltered shores of estuaries and lagoons also make them preferred locations for human settlements, which have grown into large cities in many places with associated industries and port operations.

This immediately highlights a potential conflict in the coastal zone where unique, and thus valuable, ecosystems coexist with expanding urban centres. In these contexts, the consequences of extreme events such as flooding, erosion and water quality disturbances, which can already negatively impact on the natural environment, are often further exacerbated by human activities. Thus, an Integrated Coastal Zone Management (ICZM) framework (Cicin-Sain *et al.*, 1998) is needed to manage the risks of extreme flooding, erosion and ecological events and balance blue economy initiatives focussed on development with environmental protection. With respect to marine extremes, ICZM strategies need to mitigate coastal impacts from natural extreme events and apply marine spatial planning to develop blue economy initiatives safely and sustainably. Good management requires knowledge of the present state of the coastal zone and future developments. Hence, accurate observations and predictions of events and impacts in the near and distant future are indispensable for successful sustainable coastal management.

In order to predict impacts on the natural environment and human communities we need to understand and model coastal zone processes. The complex interaction of atmospheric, marine and terrestrial processes in the coastal zone makes this a difficult task. In general, there are two relevant approaches: (1) process-based (dynamic) and (2) empirical. Process-based models imitate nature based on our understanding of the processes, within the constraints of computational resources. These process-based models must be validated against observational data, ideally measured over a range of conditions. Empirical models are data-driven and extrapolate predictions based on past experience. They are thus always limited by the amount and range of available data. Both types of modelling approaches improve with the amount of available data, thus highlighting the importance of ongoing monitoring (see Chapter 6).

This chapter reviews coastal processes that lead to extreme flooding, erosion and harmful water quality events. It highlights the challenges and recent advances in modelling such events.

Extreme flooding events

In order to protect life and property coastal planners and emergency managers require accurate estimates of flood risk. Providing reliable predictions of flood levels and duration for this purpose represents a significant challenge due to the range of complex processes that vary from beach to beach, town to town and around the global oceans. In coastal systems flooding events can be due to storm surge, waves, heavy rainfall or other processes. This section explains how flooding events occur, how short-term forecasts (real-time operational systems for emergency managers) and long-term future projections (for planning and adaptation) can be made and the challenges that remain.

Coastal regions experience a rise and fall in sea level, which can vary at timescales of hours, days, weeks, months and years, governed by astronomical tides, meteorological conditions, seismic events, local bathymetry and a range of other factors. When water levels rise above natural and man-made coastal defences, coastal flooding occurs. The gravitational pull of the moon and the sun causes regular, predictable, oscillations of the ocean surface, or astronomical tides. The magnitude of these vertical oscillations varies from centimetres (such as in the Mediterranean Sea) to over 10 metres (as observed in North-west Australia or the Bay of Fundy, Newfoundland). Tides generally occur with two highs and two lows (semi-diurnal) or a single high and low (diurnal) per day, and have a fortnightly (spring-neap) variation. Smaller, but some-times significant, variability occurs at 4.4 and 18.6 year cycles (Pugh and Woodworth, 2014).

The term 'storm surge' refers to excess sea levels above predicted tides, caused by changing atmospheric pressure (for every 1 hectopascal drop (rise) in pressure, sea level raises (lowers) by approximately 1 cm) and wind drag forces acting on the ocean surface. Storm surges can affect coastal regions over hundreds of kilometres for periods of hours to days, and are induced by extra-tropical (away from the tropics) low-pressure systems or intense tropical cyclones. These short-term fluctuations in water levels are superimposed on regional inter-annual to decadal sea level variability, such as the El Niño Southern Oscillation (ENSO) or North Atlantic Oscillation (NAO), and long-term mean sea level (MSL), which changes at timescales of centuries (Pugh and Woodworth, 2014).

Other localised processes that cause significant changes in sea level at the coast include: resonant standing waves; effects of breaking surface gravity (wind) waves through wave setup and runup; rainfall through river flooding in estuaries; seismic and meteorological tsunamis; and remotely forced coastally-trapped waves (Pugh and Woodworth, 2014).

'Extreme sea levels' can occur through any combination of these processes resulting in inundation of low-lying coastlines, loss of life and billions of dollars of damage to coastal infrastructure. Extreme flooding events often arise from a coincidence of events, such as a moderate storm surge occurring at local high tide during the peak of the spring-neap cycle; or when excessive rainfall causes rivers to rise and the release of freshwater into coastal oceans is restricted by a storm surge. For example, during September 2017, Hurricane Maria devastated Caribbean island nations and was the third costliest hurri-cane in US' history. Puerto Rico experienced wind speeds greater than 250 km/hr (just under Category 5 on the Saffir–Simpson intensity scale), with large waves and storm surges causing inundation 1–3 m above mean sea level. At the same time, intense rainfall (up to 38 inches caused by the storm), caused rivers to reach record levels around the island, compounded by flood-ing from the ocean. Maria was the most destructive hurricane to hit Puerto Rico in modern times, causing an estimated 90 billion dollars in damage and dozens of deaths, many through drowning (Pasch *et al.*, 2018).

Low-lying nations, such as Bangladesh and some island nations, are par-ticularly at risk from coastal flooding. In these nations, tropical cyclones as well as moderate surge events can cause extensive flooding and present a clear risk to coastal communities, often on a regular basis.

Over the last 150 years, global sea levels have risen on average by approxi-mately 25 cm (Church *et al.*, 2013) and it is predicted that this rise will con-tinue during the twenty-first century (and beyond) at an accelerated rate.

With rising sea levels, present-day flood levels will be exceeded increasingly frequently as, progressively, less severe storm conditions will produce equivalent water levels (Haigh *et al.*, 2014a). In some coastal regions, extreme water levels could be amplified further by changes in storminess, such as more intense tropical cyclones, although there are still significant uncertainties regarding possible future changes in tropical and extra-tropical storm activity (Church *et al.*, 2013).

In order to prepare for and respond to coastal flooding events, short-term predictions of sea level and inundation, as well as an understanding of the probability of occurrence of extreme sea levels, are required. Short-term sea level forecasts generally consist of tidal predictions based on harmonic analysis of tide gauge data, combined with dynamical forecasts derived from operational numerical storm surge and wave models (Pugh and Woodworth, 2014). These models are forced by wind and pressure data from dynamic atmospheric weather models, and thus the accuracy of sea level predictions relies on the accuracy of the weather models. Recent theoretical and computational advances, and the assimilation of observed data into these models, has dramatically improved the accuracy of forecasts with lead times of one or two weeks in some cases. Extra-tropical storm systems are generally well predicted, while the track and intensity of tropical cyclones are still difficult to forecast days in advance. To account for this uncertainty, 'ensemble' forecasts consisting of many individual model runs allow forecasters to assign probabilities of occurrence for different scenarios (Kalnay, 2002).

State-of-the-art operational forecasts, such as those that were available recently for Hurricane Maria in 2017, provide detailed predictions of wind, rain, waves, storm surge and inundation as the storm approaches, assisting emergency responders and providing warnings to residents. Without these warnings, the loss of life and destruction inflicted by Hurricane Maria would have undoubtedly been much worse.

For long-term planning purposes, historical tide gauge data, as well as empirical and statistical models, have traditionally been applied to define extreme sea levels and assign probabilities of their occurrence using extreme value theory. These probabilities are often given in terms of 'Return Periods', or 'Average Recurrence Intervals (ARI)', which describe the *average* time between events of a given magnitude. For example, 100-year ARI levels are often used for planning guidelines and suggest that the level will only be exceeded once every 100 years. However, this commonly causes confusion, as 100-year events can occur more frequently; perhaps a more useful description is that this level has a 1 in 100 or 1 per cent chance of occurring each year (the Annual Exceedance Probability or AEP).

A weakness of this approach is that data are often limited in space and time, and cannot be used to accurately determine extreme water levels in many coastal locations. As a consequence, to estimate extreme sea levels all around the coast, forecasters typically employ numerical circulation and storm-surge models forced by atmospheric weather reanalysis or synthetic tropical cyclone models. The resultant multi-decadal hindcasts of water levels over a broader area can then be analysed using extreme value theory (Haigh et al., 2014b). The results support planning and emergency management (e.g. evacuation routes) and help to map out areas at risk from inundation and erosion using dynamic models or geographic information systems (GIS).

Challenges to the accuracy of predictions (at both short and long time-scales) arise from the ability of models to resolve all of the processes and, from the limited data available, to force and validate the models. For example, tide gauges have sparse coverage and are often located in sheltered waters, not on open coasts where water levels are often higher. Other model limitations are due to: the coarse resolution of atmospheric models that generally do not predict small-scale storms correctly; the lack of waves in storm-surge models; insufficient bathymetric data; and the coincidence of rainfall and river flooding, which is generally not taken into account. Furthermore, the magnitude of future sea level rise and the changing tracks and intensities of storms are still uncertain.

Despite these challenges, the ability to forecast extreme events is rapidly improving. Advances will come from improved atmospheric and oceanic data availability; more accurate bathymetry; higher resolution and more accurate atmospheric models; faster computers; inclusion of more physical processes; and coupling between ocean circulation, wave and atmospheric models.

Extreme erosion events

Extreme erosion events occur during storms when large locally or remotely generated waves, strong currents and elevated water levels impact coastlines. Waves and currents exert shear stress on the sea bed, suspending sediment into the water column, which is subsequently transported until it settles. Under waves, water velocities oscillate periodically, which suspends sediment more efficiently than a unidirectional current.

Unprotected sandy beaches exposed to wave action are among the most dynamic coastlines. Generally, large waves that occur during extreme storms erode sand from the beach, which is deposited offshore by near-bed currents that are directed offshore. Furthermore, low atmospheric pressure and onshore winds during storms elevate the water level (storm surge) shifting the area of wave attack higher up the shore face. However, the beach response to an individual storm is not merely a function of wave height but rather controlled by a combination of wave height, wave angle, water level and pre-existing beach shape (Yates et al., 2009; Coco et al., 2014; Splinter et al., 2018). Offshore features such as sub-aqueous vegetation (see Chapter 5), sub-merged reefs, shoals and sandy barrier islands can dissipate incident waves and thus protect shorelines from wave attack during storm events. However, isolated offshore features can also focus wave energy and locally increase erosion.

Low-lying barrier islands are common features on sandy coasts around the globe (Stutz and Pilkey, 2011), and provide sheltered waterways in their lee.

However, major storms such as tropical cyclones generate energetic wave conditions and high surge levels that can lead to overwash of barrier islands. Such overwash can cause major geomorphological changes to the islands and to the shape of tidal inlets. The formation of a new breach greatly impacts the hydrodynamics, morphology and water quality in back-barrier bays by changing the location and rate of water exchange with the ocean. Several recent hurricanes in the Atlantic Ocean have impacted the US East Coast and caused the formation of new inlets between the ocean and back-barrier bays. Two examples include Hurricane Irene in 2011 and Hurricane Sandy in 2012. Hurricane Irene crossed a large system of inter-connected estuaries and lagoons in North Carolina, which generated a high storm surge and large waves within the shallow micro-tidal environment (Mulligan *et al.*, 2015). This storm event caused high-water levels, flooding and erosion of extensive areas of the low-lying coastal plain and Outer Banks barrier island system. The severe conditions led to morphological changes of the system, because the barrier islands were overwashed and breached in several locations. Hurricane Sandy generated large waves and storm surge in the Atlantic Ocean (Bennett and Mulligan, 2017). This hurricane followed a westward track that was not typical of tropical cyclones in this basin, impacting the coasts of New Jersey and New York. The nearshore region was severely impacted during the storm with extensive overwash and erosion leading to the breaching of the barrier island in three locations and large quantities of sand were moved offshore or carried into the back-barrier basin. The storm delivered strong local wind-driven storm surge in the back-barrier bay that had the largest influence on the water level fluctuations and circulations. A large proportion of ocean water entered into the bay as overwash via the barrier islands (Bennett *et al.*, 2018).

Along low-latitude coastlines, coral reefs are common features and their shallow structures typically prevent large waves from impacting upon the shoreline. Coral reefs span a range of morphologies, from being shore-attached (e.g. many Pacific Islands) to being located tens of kilometres offshore from the coast backed by extensive lagoons (e.g. Australia's Great Barrier Reef). Inside extensive lagoons, the high winds during tropical cyclones can produce locally-generated wind waves. Recent field observations from Tropical Cyclone Olwyn, which struck a fringing reef in North West Australia in 2015, indicated small coastal erosion that did not result from the ~6 m offshore waves, which were largely dissipated by the reef. The erosion was rather driven by wind waves generated within the ~2.5 km wide lagoon onshore of the reef (Cuttler *et al.*, 2018). These results and similar findings elsewhere (Mahabot *et al.*, 2016) indicate the effectiveness of reefs in providing natural protection from larger waves impacting the shoreline, and corresponding erosion. However, the effectiveness of reefs in protecting the coastline will be diminished under higher global sea levels and as a result of degradation of reefs, both of which will increase wave transmission across offshore reefs and result in increased wave energy at the shoreline (Sheppard *et al.*, 2005; Ferrario *et al.*, 2014). This future trend also increases the height of low frequency waves that are generated by non-linear interactions between different swell components. When these waves resonate on fringing reef structures they can cause substantial inundation (Péquignet *et al.*, 2009; Bosserelle *et al.*, 2015).

Erosion during extreme events can be buffered by a healthy beach and dune system that is temporarily eroded during extreme storms and subsequently restored during calmer conditions. Recovery times can be variable and strongly depend on the morphology and sediment availability in the subaqueous beach profile (Muñoz-Perez and Medina, 2010). Groups of storms, which do not leave enough recovery time, can be as detrimental as a single extreme event (Ferreira, 2005). Long-term structural erosion reduces the buffering potential of beaches and dunes that can render sandy coastlines increasingly vulnerable. Human interventions such as coastal structures that hinder alongshore sediment transport, river dams that impede the supply of sediment to the coast and sea level rise are major contributors to structural erosion. Only where sufficient sediment is available can beaches adapt to rising sea levels by accommodating more sediment (Stive, 2004). Extreme erosion events are likely to increase in frequency in many places as a result of rising sea levels, which will elevate storm surges from moderate storms to levels of present-day extreme events. Compounding this hazard is the widespread development along many coastlines with many structures built with an insufficient setback (even relative to today's sea level). There is also some evidence that storm intensity, particularly for tropical cyclones, may increase as a result of climate change (IPCC, 2014), further enhancing the risk of extreme erosion events.

Modelling coastal erosion to predict future trends requires an understanding of short-term and small-scale sediment transport processes, and how these accumulate to drive long-term and large-scale morphological changes (Gallop *et al.*, 2015; Vitousek *et al.*, 2017). Single extreme events often determine long-term beach changes because the energetic conditions can mobilise sediment grains and rocks that otherwise would not move. To predict morphology changes, process-based sediment transport models need to be coupled with hydrodynamic models (see 'Extreme flooding events' above) that calculate the forces on the sea bed and sediment transport capacity. These coupled models are computationally expensive and are thus not well-suited to evaluating shoreline changes over large temporal and spatial scales. Simplified one-line coastline models calculate erosion and accretion rates based on the alongshore sediment transport rates that are estimated from bulk parameters such as wave height and angle of wave attack but are limited to uniform coastlines. More recently, empirical Bayesian Network Approaches have been tested, in which a model is 'trained' with data on erosion and accretion patterns (Hapke and Plant, 2010; Jäger *et al.*, 2017; Plomaritis *et al.*, 2017).

Extreme water quality disturbances

Estuaries and coastal waters are typically productive ecosystems that play an important role in the local water balance, biogeochemical cycles and also in supporting biodiversity. Human activity has greatly accelerated the nutrient loadings to coastal ecosystems over the past decades, causing widespread eutrophication that remains a persistent management challenge with detrimental effects on biodiversity and the overall amenity and health of coastal waters. Worldwide coastal and estuarine waters are also being exposed to an increasingly complex suite of climatic and anthropogenic perturbations that alter their hydrology, chemistry and ecological integrity. Extreme climatic events such as droughts and floods strongly influence the delivery of freshwater and associated nutrients and organic matter to the receiving estuaries and coastal zones, often leading to large-scale ecological consequences (e.g. Paerl *et al.*, 1998; Scavia *et al.*, 2002). Extreme events such as storms and floods can also cause physical damage to the aquatic system and subsequently drive changes in their ecologic structure (Preen *et al.*, 1995).

A large diversity of extreme ecologic events occur in coastal waters, including pulsed turbidity events, hypoxia, harmful algae blooms (HABs), macrophyte blooms, jellyfish blooms, and wrack die-off (McClelland and Valiela, 1998; Anderson *et al.*, 2002; Gray *et al.*, 2002; Levin *et al.*, 2009; Pitt and Purcell, 2009; Browne *et al.*, 2015; Oldham *et al.*, 2017). Hypoxia and HABs are the most observed extreme ecologic events in coastal ecosystems and their occurrence is often associated with environmental conditions of prolonged drought, heatwaves, storms or floods (Anderson *et al.*, 2002; Gray *et al.*, 2002; Levin *et al.*, 2009). Hypoxia is more often seen in areas that suffer from low flushing rates and high organic matter loading (Bruce *et al.*, 2014). The extent and severity of oxygen depletion can be highly variable across the length of the coastal system and throughout the year, since it depends both on physical circulation patterns and biogeochemical processes such as organic matter mineralisation, photosynthesis and the sediment oxygen demand. For example, the year 2010 was an extremely dry year for the Swan River Estuary, an iconic estuary located in southwest Western Australia. The annual flow entering the system decreased to 24 gigalitres (GL) in 2010, just about 15 per cent of average annual inflow in the past decade. The extreme low freshwater inflow equated to reduced external organic matter and nutrient loading, which would be expected to decrease the risk for hypoxia. However, the prolonged drought led to an increase in severe hypoxic events, mostly in the upstream reaches. The main reason for this can be attributed to an alteration of the 'normal' movement of the salt-wedge that was pushed upstream due to reduced inflows and caused strong stratification. The strong stratification, combined with the high oxygen demand of the sediment in this region (Norlem *et al.*, 2013), lowered the oxygen concentration to persistently less than 2mg/l in the bottom layer of the water column and caused negative impacts on internal nutrient loading and higher organisms (Huang *et al.*, 2018).

Fish kills are another 'common' consequence of extreme events in coastal systems that are more often seen in areas suffering from anoxic conditions (Tweedley *et al.*, 2014) or sudden changes in the surrounding environment. Fish kills are typically triggered by abrupt events that leave little opportunity for fish to escape deleterious water quality conditions, which are often driven by flooding events. For example, a flooding event in the Murray River, the biggest inflow river to the Peel-Harvey Estuary located 50 km south of Perth in south-western Australia, triggered a fish kill event in 2017 (Valesini *et al.*, 2019). The catchment for this river is known to be a major dairy farm area in Western Australia and associated runoff is rich in organic matter. The flood washed highly labile organic matter that had accumulated over the dry summer into the Murray River. The organic matter bound large amounts of oxygen, rapidly depleting oxygen concentrations. In addition, field monitoring data revealed anoxic conditions had developed in the deep areas (> 3 m) of the Murray River prior to the event. The flood flushed anoxic water out of the deep areas and triggered the fish kill. This was not an isolated event; in the same year two other fish kill events happened (in June) after two extreme rainfall events.

HABs are one of the most notorious consequences of eutrophication in coastal systems. They pose serious threats to water quality, human health and the economy by releasing toxins and causing low oxygen kills. HABs can also cause fish kills in multiple ways such as releasing toxins, mechanical clogging of fish gills or by oxygen depletion (Landsberg, 2002). Both external loading of pollutants from anthrophonic discharge and internal loading from sediments are expected to further increase the risk of HABs in the coming decades, compounded by warming conditions associated with climate change. It is known that HABs occur when a suite of favourable environmental conditions emerges; yet determining the population dynamics of a phytoplankton species in coastal waters remains a challenge. Examples of algal blooms following extreme climate events, especially interactions between drought and floods, have been reported. In 2007, a massive bloom of the toxic dinoflagellate, *Cochlodinium polykrikoides*, arose in Chesapeake Bay coincident with a period of heavy rains that followed a drought (Mulholland *et al.*, 2009). Similarly, a bloom of toxic *Microcystis* and *Anabaena* was observed in the Swan River Estuary in late January 2000 following a storm passage that flushed over 270 GL runoff, about 5.5 times the estuary volume, into the estuary (Department of Water, 2000). The flooding events fundamentally changed the salinity distribution in the estuaries, which are critical to phytoplankton growth and succession. In addition to the salinity change, the flooding events washed a large amount of nutrients and organic matter into the estuary, creating a fresh, nutrient-rich and warm aquatic environment that favoured the growth of some specific phytoplankton species. The algae biomass peaked rapidly after the flooding event and formed a bright green scum that was visible in the areas of the upper Swan River and Canning River. In response, the rivers were closed for 12 days and the public warned not to have any contact with river water.

Interestingly, an almost identical storm occurred in February 2017 and delivered the same magnitude of freshwater runoff to the estuary, except there was no sign of algal bloom development after this event. The nutrient source was not the limiting factor for algae growth in the event of 2017 as the nitrogen and phosphorus concentrations more than tripled during the flood event compared to pre-flood conditions. Instead, cloudy post-flood weather was unfavourable for the growth of harmful algae with less light availability and lower air temperature. In addition, salinity within the estuary recovered to pre-flood conditions faster in the 2017 flooding event, which further reduced the possibility of the *Microcystis* bloom (see Figure 3.1). This highlights that water quality issues arise from a complex interaction of a myriad of environmental factors, which makes it difficult to predict these events for management purposes.

The impacts of extreme events on coastal ecosystems last from weeks to years (Wetz and Yoskowitz, 2013), and this remains a management challenge for the remediation of these systems. Although there is an increasing interest and number of reports on extreme ecologic events, our knowledge of the mechanistic linkages between physical, chemical and biological processes seems to be limited to particular events, and to be highly site and context specific. More study cases are needed to gain a generalised picture in order to better project the consequences and therefore improve the management of these systems with respect to extreme events. Ongoing monitoring of weather conditions, river inflows and the ecological state of estuaries will improve our understanding of the dominant processes at any given time (see Chapter 6 for an in-depth discussion on monitoring). Measures such as real-time observing and forecasting systems can support adaptive and proactive management of extreme ecologic events (e.g. Babin *et al.*, 2008).

Water quality models are now widely used in coastal water research and have proved to be useful tools in quantifying and predicting estuarine biogeochemical dynamics, while real-time information systems have been developed for assisting prediction and response to extreme events (e.g. Babin *et al.*, 2008; Lynch *et al.*, 2014). Field measurements of water quality are necessary for monitoring purposes, yet the data collection is usually costly and time- and labour-intensive, and therefore limited to temporal and spatial scales that adequately identify characteristic factors of extreme ecologic events in coastal systems. The nature of extreme events requires quick and accurate predictions of system changes in order to run remediation programmes effectively. Recognition of these problems has raised concerns by regulatory agencies that seek assistance from numerical water quality models. Real-time modelling has been applied in the ocean environment since the 1940s and has developed at an increasingly fast pace over the last three decades. Its application is, however, relatively new in estuarine environments due to their complex bathymetry, interaction of freshwater runoff with oceanic forcing and intensive biogeochemical processes.

In response to the challenges in the Swan Estuary mentioned above, a water quality model has been developed (Hipsey *et al.*, 2014; Huang *et al.*, 2017;

Huang *et al.*, 2018) and integrated into an online platform called the Swan-Canning Estuary Visual Observatory (SCEVO, http://swan.science.uwa.edu.au), which serves to provide real-time hydrology and water quality predictions for the system. By integrating multiple data resources from local agencies and linking to regional coastal forecasting systems (http://coastaloceanography.org), the model provides water quality hindcast for the past 5 days and forecast for the coming 5 days. The model outputs are presented online and have been used during flood events to assist investigations of hypoxia risk in the Swan–Canning Estuary and support management actions (see Figure 3.1).

Managing extreme events

Extreme events can have a range of impacts in the coastal zone, which include flooding, erosion and water quality disturbances. Management of these extreme events strives to prevent or mitigate their negative impacts and to increase the resilience of the natural and human environment. To prepare for extreme events we not only need to understand the potential impacts of an individual extreme event, but also the cumulative impacts of multiple events. This highlights the importance of good monitoring strategies that can capture the single and cumulative impacts of past events (as discussed in detail in Chapter 6). Short-term forecasts produced by real-time systems are essential to respond to imminent events, for example, to evacuate people from areas at risk. To adapt to changing extreme events due to climate change, projections of future long-term trends are indispensable. However, even if perfect predictions were available, sustainable, science-based planning and management is needed to guide coastal communities into an uncertain future of a changing environment with minimal human and economic costs.

Modelling tools can predict the outcomes of various management strategies and are thus helpful in decision-making processes. Adequate impact modelling tools for coastal managers should combine different forms of extremes to address compounding effects related to a single event. As an example, in 2011 Hurricane Irene impacted the coast of North Carolina in multiple ways. In addition to the storm surge, strong winds and large waves (see 'Extreme erosion events' above), which caused flooding and erosion, the hurricane also delivered intense rainfall. The flooding related to precipitation persisted for weeks while the wind-driven storm surge lasted for hours. Brown *et al.* (2014) used a numerical model to investigate the ecological impact of the flooding. The model simulated freshwater river discharge into a brackish estuary and tracked the transport of dissolved organic carbon. Their work indicated that the increased flux of freshwater progressed down the estuary as a shallow near-surface plume that impacted the estuarine residence time and reduced phytoplankton production.

There are multiple strategies to prevent or mitigate the impact of extreme events such as spatial planning and structural and nature-based solutions (the latter are discussed in detail in Chapter 5). It should be noted that choosing

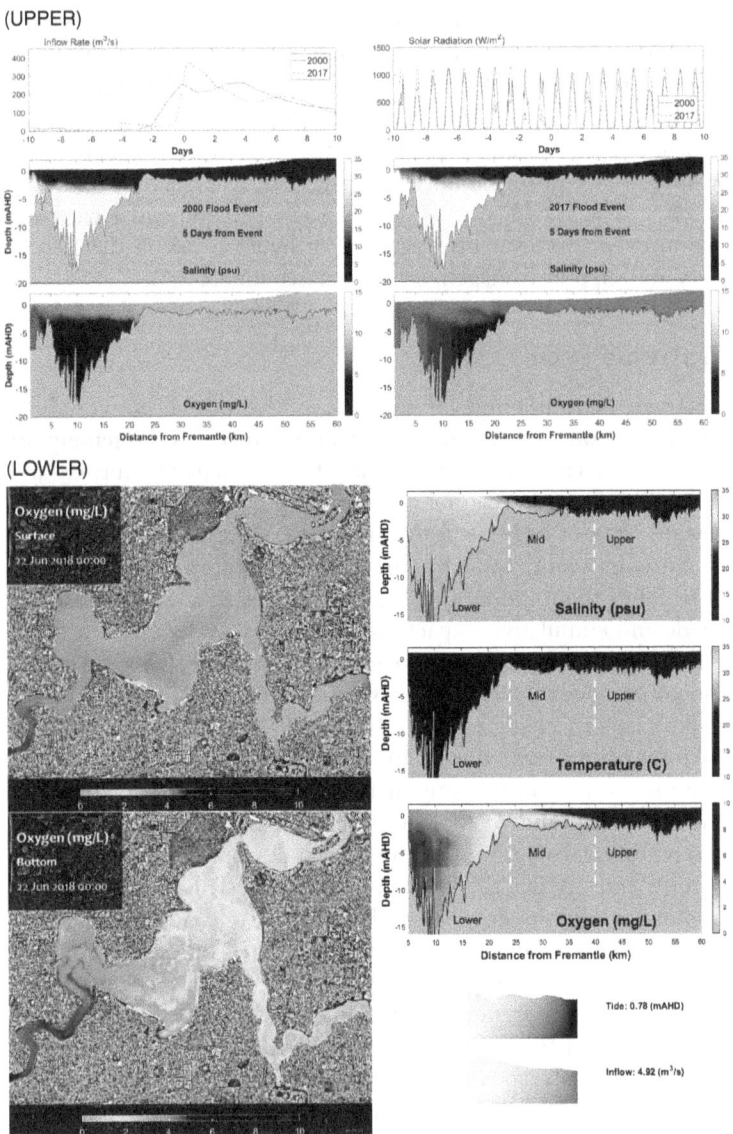

Figure 3.1 (UPPER) Modelled longitudinal distributions of salinity and oxygen concentration distribution along the Swan River. (LOWER) Screen snapshot of the SCEVO real-time system.

Source: figure created by Peisheng Huang.

Note
(UPPER) The modelling results took place 5 days after the 2000 and 2017 flooding events. The first row shows the inflow rates and solar radiation during these two events. (LOWER) Shows the predicted surface and bottom oxygen concentrations in the Swan River, along with the longitudinal distribution of salinity, temperature and oxygen concentration and real-time tide and inflow information.

the right strategy to cope with extreme events in the coastal zone is not purely a scientific and engineering task but needs to acknowledge community preferences and expectations (see Chapter 4). Decentralising coastal management, for example, can give ownership to the local community and ensure that management strategies are consistent with local preferences and customs (see Chapter 9).

Conclusions

Extreme storm events impact upon the coastal zone through flooding, erosion and ecological disturbances with negative consequences for the environment, coastal communities and economic activities. The development of successful management solutions to prevent or mitigate negative impacts requires collaboration across multiple disciplines including scientists, engineers and social scientists. It is their task to design modelling tools that can predict the impacts of extreme events now and in the future. Such predictions are valuable for government decision-makers as well as for industry and communities with interests in the coastal zone. For the development of these tools and the ongoing evaluation of management strategies, long-term monitoring and data collection are indispensable. Through ICZM, human communities and economic activities can develop safely and sustainably – even where extreme events are likely to intensify due to climate change.

References

Anderson, D. M., Glibert, P. M. and Burkholder, J. M. (2002). 'Harmful algal blooms and eutrophication: Nutrient sources, composition, and consequences'. *Estuaries*, vol 25, no 4, pp. 704–726, doi:10.1007/BF02804901.

Babin, M., Roesler, C. S. and Cullen, J. J. (eds.) (2008). *Real-time Coastal Observing Systems for Marine Ecosystem Dynamics and Harmful Algal Blooms: Theory, Instrumentation and Modelling*. Paris: UNESCO Publishing.

Bennett, V. C. C. and Mulligan, R. P. (2017). 'Evaluation of surface wind fields for prediction of directional ocean wave spectra during hurricane sandy'. *Coastal Engineering*, vol 125, pp. 1–15, doi:10.1016/j.coastaleng.2017.04.003.

Bennett, V. C. C., Mulligan, R. P. and Hapke, C. J. (2018). 'A numerical model investigation of the impacts of Hurricane Sandy on water level variability in Great South Bay, New York'. *Continental Shelf Research*, vol 161, pp. 1–11, doi:10.1016/j.csr.2018.04.003.

Bosserelle, C., Kruger, J., Movono, M. and Reddy, S. (2015). 'Wave inundation on the Coral Coast of Fiji'. In Australasian Coasts & Ports Conference 2015: 22nd Australasian Coastal and Ocean Engineering Conference and the 15th Australasian Port and Harbour Conference. Auckland, New Zealand: Engineers Australia and IPENZ, pp. 96–101.

Brown, M. M., Mulligan, R. P. and Miller, R. L. (2014). 'Modeling the transport of freshwater and dissolved organic carbon in the Neuse River Estuary, NC, USA following Hurricane Irene (2011)'. *Estuarine, Coastal and Shelf Science*, vol 139, pp. 148–158.

Browne, N. K., Tay, J. and Todd, P. A. (2015). 'Recreating pulsed turbidity events to determine coral–sediment thresholds for active management'. *Journal of Experimental Marine Biology and Ecology*, vol 466, pp. 98–109.

Bruce, L. C., Cook, P. L. M., Teakle, I. and Hipsey, M. R. (2014). 'Hydrodynamic controls on oxygen dynamics in a riverine salt wedge estuary, the Yarra River estuary, Australia'. *Hydrology and Earth System Sciences*, vol 18, no 4, pp. 1397–1411.

Church, J. A., Clark, P. U., Cazenave, A., Gregory, J. M., Jevrejeva, S., Levermann, A., Merrifeld, M. A., Milne, G. A., Nerem, R. S., Nunn, P. D., Payne, A. J., Pfeffer, W. T., Stammer, D. and Unnikrishnan, A. S. (2013). 'Sea-level rise by 2100'. *Science*, vol 342, no 6165, p. 1445, doi:10.1126/science.342.6165.1445.

Cicin-Sain, B., Knecht, R. W., Jang, D., Knecht, R. and Fisk, G. W. (1998). *Integrated Coastal and Ocean Management: Concepts and Practices*. Washington, DC: Island Press.

Coco, G., Senechal, N., Rejas, A., Bryan, K. R., Capo, S., Parisot, J. P., Brown, J. A. and MacMahan, J. H. M. (2014). 'Beach response to a sequence of extreme storms'. *Geomorphology*, vol 204, pp. 493–501, doi:10.1016/j.geomorph.2013.08.028.

Cuttler, M. V. W., Hansen, J. E., Lowe, R. J. and Drost, E. J. F. (2018). 'Response of a fringing reef coastline to the direct impact of a tropical cyclone'. *Limnology and Oceanography Letters*, vol 3, no 2, pp. 31–38.

Department of Water. (2000). '"Summer surprise" – The Swan River blue–green algal bloom'. *River Science*, vol 2.

Ferrario, F., Beck, M. W., Storlazzi, C. D., Micheli, F., Shepard, C. C. and Airoldi, L. (2014). 'The effectiveness of coral reefs for coastal hazard risk reduction and adaptation'. *Nature Communications*, vol 5, doi:10.1038/ncomms4794.

Ferreira, O. (2005). 'Storm groups versus extreme single storms: Predicted erosion and management consequences'. *Journal of Coastal Research*, vol SI42, pp. 221–227.

Gallop, S. L., Collins, M., Pattiaratchi, C. B., Eliot, M. J., Bosserelle, C., Ghisalberti, M., Collins, L. B., Eliot, I., Erftemeijer, P. L. A., Larcombe, P., Marigómez, I., Stul, T. and White, D. J. (2015). 'Challenges in transferring knowledge between scales in coastal sediment dynamics'. *Frontiers in Marine Science*, vol 2, no 82, pp. 1–7, doi:10.3389/fmars.2015.00082.

Gray, J. S., Wu, R. S.-S. and Or, Y. Y. (2002). 'Effects of hypoxia and organic enrichment on the coastal marine environment'. *Marine Ecology Progress Series*, vol 238, pp. 249–279.

Haigh, I., MacPherson, L., Mason, M., Wijeratne, E. M. S., Pattiaratchi, C., Crompton, R. and George, S. (2014a). 'Estimating present day extreme water level exceedance probabilities around the coastline of Australia: Tropical cyclone-induced storm surges'. *Climate Dynamics*, vol 42, no 1–2, pp. 139–157, doi:10.1007/s00382-012-1653-0.

Haigh, I., Wijeratne, E. M. S., MacPherson, L., Pattiaratchi, C., Mason, M., Crompton, R. and George, S. (2014b). 'Estimating present day extreme water level exceedance probabilities around the coastline of Australia: Tides, extra-tropical storm surges and mean sea level'. *Climate Dynamics*, vol 42, no 1–2, pp. 121–138, doi:10.1007/s00382-012-1652-1.

Hapke, C. and Plant, N. (2010). 'Predicting coastal cliff erosion using a Bayesian probabilistic model'. *Marine Geology*, vol 278, no 1–4, pp. 140–149, doi:10.1016/j.margeo.2010.10.001.

Hipsey, M. R., Kilminster, K., Busch, B. D., Bruce, L. C. and Larsen, S. (2014). 'Modelling oxygen dynamics in the Upper Swan estuary and Canning Weir Pool'. AED Report #R25, The University of Western Australia, Perth.

Huang, P., Hipsey, M. R. and Busch, B. (2017). 'The Swan-Canning Estuary Response Model (SCERM) v2: Model validation, monitoring data assessment and real-time operation'. AED Report #R34, The University of Western Australia, Perth.

Huang, P., Kilminster, K., Larsen, S. and Hipsey, M. R. (2018). 'Assessing artificial oxygenation in a riverine salt-wedge estuary with a three-dimensional finite-volume model'. *Ecological Engineering*, vol 118, pp. 111–125.

IPCC (2014). 'Climate change 2014: Synthesis report. Contribution of working groups I, II and III to the fifth assessment report of the Intergovernmental Panel on Climate Change' [Core writing team, R. K. Pachauri and L. A. Meyer (eds.)]. IPCC, Geneva.

Jäger, W. S., Christie, E. K., Hanea, A. M., den Heijer, C. and Spencer, T. (2017). 'A Bayesian network approach for coastal risk analysis and decision making'. *Coastal Engineering*, vol 134, pp. 48–61, doi:10.1016/j.coastaleng.2017.05.004.

Kalnay, E. (2002). *Atmospheric Modeling, Data Assimilation and Predictability*. Cambridge: Cambridge University Press.

Landsberg, J. H. (2002). 'The effects of harmful algal blooms on aquatic organisms'. *Reviews in Fisheries Science*, vol 10, no 2, pp. 113–390.

Levin, L. A., Ekau, W., Gooday, A. J., Jorissen, F., Middelburg, J. J., Naqvi, S. W. A., Neira, C., Rabalais, N. N. and Zhang, J. (2009). 'Effects of natural and human-induced hypoxia on coastal benthos'. *Biogeosciences*, vol 6, pp. 2063–2098.

Lynch, T. P., Morello, E. B., Evans, K., Richardson, A. J., Rochester, W., Steinberg, C. R., Roughan, M., Thompson, P., Middleton, J. F. and Feng, M. (2014). 'IMOS National Reference Stations: A continental-wide physical, chemical and biological coastal observing system'. *PLoS ONE*, vol 9, no 12, doi:10.1371/journal. pone.0113652.

McClelland, J. W. and Valiela, I. (1998). 'Linking nitrogen in estuarine producers to land-derived sources'. *Limnology and Oceanography*, vol 43, no 4, pp. 577–585.

Mahabot, M.-M., Pennober, G., Suanez, S., Troadec, R. and Delacourt, C. (2016). 'Effect of tropical cyclones on short-term evolution of carbonate sandy beaches on Reunion Island, Indian Ocean'. *Journal of Coastal Research*, vol 33, no 4, pp. 839–853.

Mulholland, M. R., Morse, R. E., Boneillo, G. E., Bernhardt, P. W., Filippino, K. C., Procise, L. A., Blanco-Garcia, J. L., Marshall, H. G., Egerton, T. A. and Hunley, W. S. (2009). 'Understanding causes and impacts of the dinoflagellate, Cochlodinium polykrikoides, blooms in the Chesapeake Bay'. *Estuaries and Coasts*, vol 32, no 4, pp. 734–747.

Mulligan, R. P., Walsh, J. P. and Wadman, H. M. (2015). 'Storm surge and surface waves in a shallow lagoonal estuary during the crossing of a hurricane'. *Journal of Waterway, Port, Coastal, and Ocean Engineering*, vol 141, no 4, doi:10.1061/(ASCE) WW.1943-5460.0000260.

Muñoz-Perez, J. J. and Medina, R. (2010). 'Comparison of long-, medium- and short-term variations of beach profiles with and without submerged geological control'. *Coastal Engineering*, vol 57, no 3, pp. 241–251. doi:10.1016/j.coastal eng.2009.09.011.

Norlem, M., Paraska, D. and Hipsey, M. R. (2013). 'Sediment-water oxygen and nutrient fluxes in a hypoxic estuary.' Paper presented at the MODSIM2013, 20th International Congress on Modelling and Simulation, Adelaide Convention Centre, Adelaide, 1–6 December 2013.

Oldham, C., McMahon, K., Hipsey, M. R., Huang, K., Huang, P. and Lavery, P. S. (2017). 'The impact of marine wrack degradation on the water quality of Jurien Bay Boat Harbour'. Report to Western Australian Department of Transport, Project 306815, The University of Western Australia, Perth.

Paerl, H. W., Pinckney, J. L., Fear, J. M. and Peierls, B. L. (1998). 'Ecosystem responses to internal and watershed organic matter loading: Consequences for hypoxia in the eutrophying Neuse River Estuary, North Carolina, USA'. *Marine Ecology Progress Series*, vol 166, pp. 17–25.

Pasch, R. J., Penny, A. B. and Berg, R. (2018). 'Hurricane Maria tropical cyclone report' (AL152017). National Oceanic and Atmospheric Administration National Hurricane Center, Miami, FL.

Péquignet, A. C. N., Becker, J. M., Merrifield, M. A. and Aucan, J. (2009). 'Forcing of resonant modes on a fringing reef during tropical storm Man-Yi'. *Geophysical Research Letters*, vol 36, no 3, doi:10.1029/2008GL036259.

Pitt, K. and Purcell, J. (2009). 'Jellyfish blooms: Causes, consequences, and recent advances'. *Hydrobiologia*, vol 616, pp. 1–5. doi:10.1007/s10750-008-9599-2.

Plomaritis, T. A., Costas, S. and Ferreira, Ó. (2017). 'Use of a Bayesian Network for coastal hazards, impact and disaster risk reduction assessment at a coastal barrier (Ria Formosa, Portugal)'. *Coastal Engineering*, vol 134, pp. 134–147. doi:10.1016/j. coastaleng.2017.07.003.

Preen, A. R., Long, W. J. L. and Coles, R. G. (1995). 'Flood and cyclone related loss, and partial recovery, of more than 1000 km^2 of seagrass in Hervey Bay, Queensland, Australia'. *Aquatic Botany*, vol 52, no 1–2, pp. 3–17.

Pugh, D. T. and Woodworth, P. L. (2014). *Sea-Level Science: Understanding Tides, Surges, Tsunamis and Mean Sea-Level Changes*. Cambridge: Cambridge University Press.

Scavia, D., Field, J. C., Boesch, D. F., Buddemeier, R. W., Burkett, V., Cayan, D. R., Fogarty, M., Harwell, M. A., Howarth, R. W. and Mason, C. (2002). 'Climate change impacts on US coastal and marine ecosystems'. *Estuaries*, vol 25, no 2, pp. 149–164.

Sheppard, C., Dixon, D. J., Gourlay, M., Sheppard, A. and Payet, R. (2005). 'Coral mortality increases wave energy reaching shores protected by reef flats: Examples from the Seychelles'. *Estuarine, Coastal and Shelf Science*, vol 64, no 2, pp. 223–234. doi:https://doi.org/10.1016/j.ecss.2005.02.016.

Small, C. and Nicholls, R. J. (2003). 'A global analysis of human settlement in coastal zones'. *Journal of Coastal Research*, vol 19, no 3, pp. 584–599.

Splinter, K. D., Kearney, E. T. and Turner, I. L. (2018). 'Drivers of alongshore variable dune erosion during a storm event: Observations and modelling'. *Coastal Engineering*, vol 131, pp. 31–41, doi:https://doi.org/10.1016/j.coastaleng.2017. 10.011.

Stive, M. J. F. (2004). 'How important is global warming for coastal erosion?'. *Climatic Change*, vol 64, no 1, pp. 27–39. doi:10.1023/B:CLIM.0000024785.91858.1d.

Stutz, M. L. and Pilkey, O. H. (2011). 'Open-ocean barrier islands: Global influence of climatic, oceanographic, and depositional settings'. *Journal of Coastal Research*, vol 27, no 2, pp. 207–222.

Tweedley, J. R., Keleher, J., Cottingham, A., Beatty, S. J. and Lymbery, A. J. (2014). 'The fish fauna of the Vasse-Wonnerup and the impact of a substantial fish kill event'. Centre for Fish and Fisheries Research, Murdoch University, Perth.

Valesini, F. J., Hallett, C. S., Hipsey, M. R., Kilminster, K. L., Huang, P. and Hennig, K. (2019). 'Peel-Harvey Estuary', in E. Wolanski, J. W. Day, M. Elliott and R. Ramesh (eds.), *Coasts and Estuaries: The Future* (1st edn.). Elsevier (in press).

Vitousek, S., Barnard, P. L. and Limber, P. (2017). 'Can beaches survive climate change?'. *Journal of Geophysical Research: Earth Surface*, vol 122, no 4, pp. 1060–1067.

Wetz, M. S. and Yoskowitz, D. W. (2013). 'An "extreme" future for estuaries? Effects of extreme climatic events on estuarine water quality and ecology'. *Marine Pollution Bulletin*, vol 69, no 1–2, pp. 7–18.

Yates, M. L., Guza, R. T. and O'Reilly, W. C. (2009). 'Equilibrium shoreline response: Observations and modeling'. *Journal of Geophysical Research: Oceans*, vol 114, no C9, doi:10.1029/2009JC005359.

4 Community values and preferences for coastal hazard interventions

Abbie A. Rogers and Michael P. Burton

Introduction

As the marine environment becomes more extreme, coastal communities will face an increasing threat from coastal hazards. Common hazards and events that are experienced globally include erosion, inundation, storms, hurricanes and tsunamis (Berz *et al.*, 2001). Exacerbated by the impacts of climate change, these hazards are growing in frequency, intensity and severity (Wong *et al.*, 2014). It is estimated that, globally, approximately 625 million people lived in the low-elevation coastal zone in 2000, and that this will grow to 880 million in 2030, and exceed 1 billion by 2060 (Neumann *et al.*, 2015). This means that these hazard events are likely to have substantial impacts on the welfare of a large number of people.

With such long-reaching impacts, and competition for resources to manage other hazards and environmental challenges, it will be increasingly important for governments to be able to effectively prioritise investment in coastal hazards (ERG, 2016). It will be necessary to assess whether adaptation is more cost-effective than the clean-up bill after damage, and which adaptation options provide the greatest improvements in welfare. To be able to deliver the maximum benefits to affected communities, policy makers need to consider the welfare of the multiple stakeholder groups affected. There are different, often conflicting, uses of the coastal environment to manage, including recreational, commercial and environmental uses. Effective hazard-management policy will require an understanding of what adaptation options are technically feasible, and how communities value the outcomes of those options.

Understanding the technical feasibility of different adaptation options requires expert input, for estimating both risks and consequences, and also the costs of adaptation activities. But the implementation costs are only part of the picture: measuring the benefits of different adaptation actions is also important. There may be cases where a particular adaptation project is more expensive than its alternatives, but it delivers far greater benefits with respect to community welfare. To measure the net benefits of hazard adaptation options, decision makers need to be able to quantify the community values for the outcomes associated with those options. These values could be in respect of tangible outcomes, such as the economic benefits of an adaptation project, or intangible outcomes, such as the environmental and social benefits.

This chapter discusses the various ways in which community values and preferences can contribute towards effective coastal hazard management policy. We discuss the range of approaches that are available to measure community values in relation to coastal hazard management, focussing on the values for social and environmental outcomes. These approaches are aimed at understanding which, from a given set of potential policy options, are the most preferred. An economic approach to how values for intangible outcomes can be monetised is used as an illustration. This shows how it is possible to trade off social, environmental and economic outcomes in prioritising investment in hazard adaptation.

Measuring community preferences

Coastal hazards and their management can affect a wide range of coastal assets that are valued by human communities. These assets include market assets, for example businesses dependent on the coast such as fisheries and tourism enterprises, which have economic value and can be measured by market-based instruments. Kirkpatrick (2011, 2012) provides an overview of coastal assets, and different approaches to valuing them. Non-market assets, which have social and environmental value, can also be affected by coastal hazards, and require a different set of approaches to quantify the values. Here, we focus on how non-market values can be measured, noting a variety of approaches ranging from qualitative to quantitative, and non-economic to economic. An emphasis is placed on discussing economic non-market valuation approaches, which have the ability to quantify these values in a manner that enables trade-offs to be made with the economic costs and benefits of coastal hazard investments.

Values and assets affected by coastal hazards

A range of non-market coastal assets may be valued by the community. The Department of Climate Change (2009) provided an initial assessment of the Australian coastal environments and built infrastructure and industry that may be at risk from climate change, as a precursor to identifying appropriate adaptation responses. The assets at risk include coastal infrastructure, such as recreational facilities, shelter, exercise and play equipment, jetties, car parking and coastal drives, cycle and walkways, surf life-saving clubs and grassed foreshore reserves. Similar assets at risk have been identified in assessments undertaken internationally (e.g. Polomé, 2002; Policy Research Corporation, 2009; Melillo *et al.*, 2014).

These types of assets generally have what is known as 'use value'; that is, individuals who use the asset gain a benefit from its use. Other assets that might be valued by the community include the natural environment such as beaches, dunes and marine and coastal ecosystems, and also locations, buildings or other infrastructure that are recognised for their cultural heritage. These assets can also have use value, but often have 'passive use value', or non-use value, associated with them in addition. That is, individuals may value the existence of an asset even if they do not directly use it. When measuring values, it is important to recognise which types of values are relevant, and who in the community might hold those values. For example, a fishing jetty might only be valued by the recreational fishers who use it, while a threatened coastal ecosystem might be valued by a regional or even a national population if it is important enough.

Different methods are appropriate for measuring different types of value – some can measure both use and non-use values of assets, while others only measure use-related values – and for capturing those values in different ways, including economic and non-economic approaches.

Non-economic approaches

There are a range of non-economic approaches that can be used to identify values. Christie *et al.* (2012) provide an overview of alternative methods, including quantitative and qualitative approaches which take the form of consultative methods, deliberative and participatory methods and review-based methods:

- Consultative methods include the application of questionnaire or interview formats as a structured process to understand people's perceptions of a particular issue. Both formats are typically applied to individuals in a sample of respondents. In-depth interviews typically gather qualitative information, while questionnaires often collect quantitative data (e.g. through scoring systems such as ratings and rankings of items).
- Deliberative and participatory approaches are group-based formats focussed on understanding the relationships of people in the group with the asset being assessed. This set of approaches includes: focus groups, which are small group discussion sessions about an issue, and may involve semi-structured question scripts; citizen juries, where participants review evidence and then make judgements on the future of an asset, as in a court-like procedure; Q-methodology, which asks individuals to sort and rank items (e.g. a series of statements about the qualities or importance of an asset); Delphi surveys, which involve a multi-round process where individuals are consulted about an issue, then responses are collated, summarised and made available to the group and the individuals (or group collectively) are consulted again (and the process may be repeated); and public participatory geographic information systems (GIS), which use maps to ask respondents to identify which locations may hold values for them.

- Systematic reviews involve the assessment of existing scientific evidence on the potential outcomes resulting from different interventions. The protocols for and outcomes of the review should be peer-assessed pre- and post-review, respectively.

While quantitative analyses are generally preferred for the generation of data suitable for inclusion in decision support tools, qualitative approaches can be particularly useful as a first step to identifying the scope of issues relevant to a population in relation to the affected asset. This approach can then be followed by a quantitative assessment of values across a representative sample of the relevant community.

Non-economic approaches can generally be used to assess use and non-use values. With respect to prioritising investments, the disadvantage of non-economic approaches is that the values measured cannot be used in a directly comparable manner in trade-off decisions with other market-based costs and benefits.

Economic approaches – non-market valuation

Non-market valuation measures values in monetary or financial equivalent terms. This means that dollar estimates are provided, which can be used to compare directly against other monetary costs and benefits associated with coastal hazard management. Values are estimated as an individual's willingness to pay for improvements or changes in the quantity or quality of an asset. The estimates can then be aggregated to the relevant population to indicate the overall value of the asset.

Non-market values can be used instrumentally in policy and decision making, for example, by using the values in damage assessment and compensation claims (e.g. Carson *et al.*, 2003; Bishop *et al.*, 2017) in environmental policy management (e.g. see review by Rogers *et al.*, 2015 for applications in Australia) or in benefit–cost analyses (Pandit *et al.*, 2015; ERG, 2016). There are two general forms of non-market valuation: revealed and stated preference methods.

Revealed preference methods

Revealed preference methods focus on observations of people's behaviour, for example, how they use an asset or how they make purchases in markets that are associated with the asset being valued. These observations are used to infer how much people are willing to pay for the asset. These methods are limited to situations where it is possible to measure use values. The travel-cost method and hedonic pricing method are two of the most commonly applied revealed preference approaches.

Recreation values are often measured through travel-cost applications. Information is collected about the costs of making a trip to a site; for example, fuel costs, entrance fees and purchases made at the site or on the way to and from it (e.g. food, souvenirs). Information is also collected about non-monetary expenses, including the implicit time cost of travel (Hanley and Barbier, 2009). This information is used to estimate how much people are willing to pay per visit to a site. We can then explore how willingness to pay varies by site, based on the different characteristics at each site.

Hedonic pricing analyses investigate how an asset's characteristics relate to the market value of the asset (Perman *et al.*, 1999; Hanley and Barbier, 2009). Property markets, for example, are often analysed to estimate the values for scenic coastal views or proximity to a beach. By examining the premiums that are paid for housing with these non-market characteristics we can infer people's willingness to pay for the coastal asset.

Stated preference methods

Stated preference methods use survey-based instruments that are used to ask individuals how much they are willing to pay to achieve an outcome or about their preferences for making trade-offs between different outcomes. Stated preference methods have the advantage of being able to measure both use and non-use values (Bateman *et al.*, 2002). It is the only approach capable of assigning monetary estimates to non-use values. Contingent valuation, contingent behaviour and discrete choice experiments are common stated preference methods (Bateman *et al.*, 2002; Johnston *et al.*, 2017).

Contingent valuation asks individuals how much they would be willing to pay to implement a positive change or prevent a negative one in order to estimate the value of assets (Hanley and Barbier, 2009). For example, individuals might be asked how much they are willing to pay to implement a beach nourishment programme to alleviate the impacts of erosion and maintain an area for recreation.

Like contingent valuation, contingent behaviour also measures people's demand for change in an asset but not through willingness to pay. For example, it might focus on changes in visitation rates relative to proposed changes in the quality or quantity of an asset (Bateman *et al.*, 2002). This enables us to understand whether, for example, individuals might make fewer or more trips to the beach, in different scenarios which each lead to altered recreational opportunities or changes in amenity through various actions such as leaving erosion unmanaged, implementing beach nourishment programmes or constructing a seawall.

Discrete choice experiments estimate how individuals make trade-offs between a set of outcomes that are defined by different features, or attributes, of an asset (Bennett and Blamey, 2001). Respondents are given a sequence of hypothetical choice scenarios, where each scenario is comprised of a number of options. The options describe, for example, different coastal hazard management programmes in terms of their attributes. The level of each attribute, in terms of the marginal change in quantity or quality, varies across the different options. One of the attributes that is usually included in the trade-off scenario is a cost which is used to calculate willingness to pay. For example, we could estimate how much people are willing to pay for protecting different lengths of foreshore infrastructure with a seawall relative to having different lengths of sandy beaches left available for recreation.

Applications of non-market valuation to coastal values

There are thousands of published environmental and social value applications using the approaches discussed above (Carson, 2011). These include many international applications to coastal assets, including some focussed on hazard contexts. A selection of relevant studies is discussed here to illustrate how different methods have been applied and the value measurements they provide.

Matthews *et al.* (2017) evaluated coastal erosion management using dunes or seawalls on the Coromandal Peninsular in New Zealand, representing the changes using virtual video representations in a choice experiment. They found that dune restoration was valued over the use of seawalls, although there was considerable heterogeneity in the sample. New Zealand residents who had visited the Peninsular in the past year were willing to pay up to NZ$134 annually per 800 m of restored dune.

Huang *et al.* (2007) considered the multiple effects of beach erosion control in a choice experiment of coastal areas in New Hampshire and Maine, United States. They concluded that the economic benefit of erosion control can be grossly overstated if just the positive benefits of beach length are evaluated, while ignoring the potential negative consequences of interventions, such as disturbance to wildlife or impacts on neighbouring beaches.

Dribek and Voltaire (2017) used the contingent valuation method to evaluate willingness to pay to protect beaches in Djerba, Tunisia, using a novel erosion control technique that minimises the impact on natural coastal processes. Using a sample of local and tourist respondents from Djerba Island, they found similar, positive values for both groups, with aggregate willingness to pay estimated in excess of 5 million euro.

Alves *et al.* (2015) found a very low level of willingness to pay for beach maintenance in Cadiz, Spain, with over 87 per cent of respondents giving a zero value in their contingent valuation study. This was despite a high level of awareness of coastal erosion among the beach users surveyed.

Kontogianni *et al.* (2014) considered the willingness to pay, to manage the exposure of beachrock due to coastal erosion, of tourists at two beaches in Greece. They found that 50 per cent would be willing to pay to manage this effect and, of those who had a positive value, the average annual willingness to pay was between 13.2 and 16.4 euro.

Windle and Rolfe (2014) addressed the values associated with provision of beach facilities and the management of coastal erosion in southeast

Queensland using discrete choice experiment and contingent valuation approaches. A sample of Brisbane respondents, including users and non-users of local beaches, were willing to pay AUD$101 per household per year (for 5 years) to reduce erosion impacts over 75 km of beach ($1.35/km). Respondents were willing to pay, per household per year (for 3 years), AUD$26 for toilet facilities to be provided at beaches, AUD$20 for monthly beach cleaning and AUD$36 for lifeguards to be present at peak periods during weekends and holidays.

While the studies above focus specifically on values associated with coastal hazard management, there are many relatable studies that measure values for the types of coastal assets that could be affected by hazards, with the valuation conducted in other contexts. For example, Peters and Hawkins (2009) reviewed 18 international studies that estimated willingness to pay for entry fees to marine parks, while Londoño and Johnston (2012) used a meta-analysis of 27 valuation studies to review willingness to pay for tropical reef recreation. Torres and Hanley (2016) reviewed 196 international studies on the valuation of coastal and marine ecosystem services. They found that coastal and marine ecosystems have high recreational benefits and that these are positively correlated with the quality of, and high magnitudes of willingness to pay for, ecosystem protection. Non-use values including those related to cultural services and biodiversity were also important.

Pilot study: quantifying community values for coastal hazard management in Western Australia

In an attempt to address the increasing threat of coastal hazards on assets in Western Australia, the State Government created the Coastal Hazard Risk Management and Adaptation Planning (CHRMAP) Guidelines (Western Australian Planning Commission, 2014). The Guidelines support the implementation of *State Planning Policy No. 2.6 – State Coastal Planning Policy* by assisting planners and coastal land managers in developing and implementing effective CHRMAP. Figure 4.1 illustrates the required steps to be taken in the CHRMAP process.

The primary hazards threatening coastal assets in Western Australia are erosion and inundation, and resulting damage to assets (Western Australian Planning Commission, 2014). The management hierarchy for coastal hazards, as set out by the CHRMAP Guidelines (Western Australian Planning Commission, 2014), is:

- avoid, e.g. by not building assets where they can be affected;
- planned or managed retreat, where losses are accepted and moveable assets relocated;
- accommodate, e.g. design assets to withstand the impacts of hazards;
- protect, e.g. construction of infrastructure to protect assets from hazard impacts.

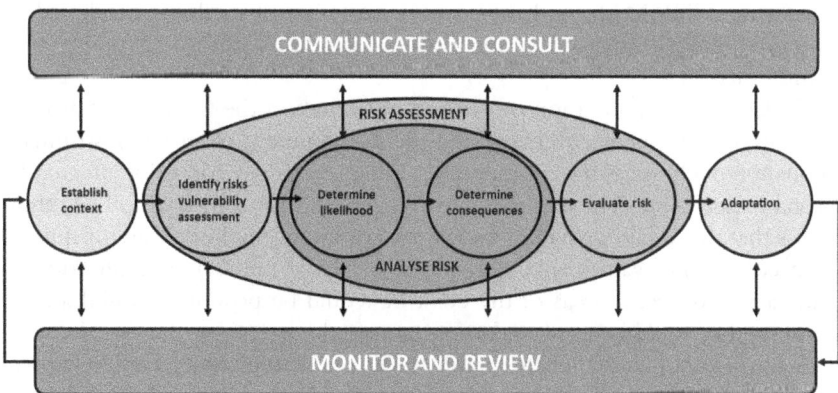

Figure 4.1 Risk management and adaptation process flowchart.
Source: adapted from Western Australian Planning Commission 2014, p. 5.

Protection is a last resort option based on the beneficiary-pays principle to ensure a sustainable approach that minimises the risk to public funds (Western Australian Government, 2017). Common forms of protection against erosion and inundation include 'hard' engineering works such as seawalls, groynes and offshore breakwaters, as well as 'soft' measures, for example, beach nourishment.

The changes in condition of the asset, due to either the hazard processes or their management, result in changes to the values held for the assets. Given the length of coastline in Western Australia, ranging from populated urban areas to remote locations, there is a wide range of assets that can potentially be affected by coastal hazards or their management, including those listed in the section 'Values and assets affected by coastal hazards' (see p. 49). Understanding community values for these assets is relevant to the CHRMAP process at several stages (see Figure 4.1):

1   The risk identification phase, where an understanding of which assets are considered important is necessary to know which assets require a risk assessment.
2   The risk evaluation phase, where information about the community's tolerable levels of risks and preferred adaptation options is required.
3   Deciding on the appropriate adaptation approach by comparing risk treatment options, where community values are needed to include in decisions about which adaptation options are the most effective, along with other information about the technical feasibilities of the different options and their implementation costs. Here, it is particularly useful if the community values are quantified in monetary-equivalent terms such that they can be included in benefit–cost analyses.

There is no formal guidance on what form community engagement could take in the CHRMAP, and there is a wide range of methods employed by local governments engaged in this process, including forums and online community surveys.

To facilitate the inclusion of community values in the CHRMAP process, the State Government commissioned the development of survey instruments to quantitatively measure community willingness to pay for assets affected by coastal hazards and their management (Rogers and Burton, 2018). The intention is that this could provide a basis for a consistent methodology for quantifying community values and, if a sufficiently large number of applications is made across different areas of the coast, it would be possible to build a bank of values that could be used for benefit transfer.[1]

A discrete choice experiment and travel cost survey were developed and piloted for Cottesloe Beach, Western Australia. The purpose of the pilot was to test the instrument that was developed, and to have results available to illustrate the approach to other managers. As well as the survey instrument itself, the project developed guidance for managers on how to adapt the survey instruments to different coastal locations.

The survey comprised two means by which to assess community values: a stated preference discrete choice experiment that identifies the values that people have as a result of degradation in coastal assets; and a travel cost component that enables identification of the values associated with beach visitation, through revealed preferences.

The discrete choice experiment was designed to estimate the values of five attributes that were representative of the types of coastal assets that are: (1) common across coastal locations in Western Australia; (2) likely to be impacted by coastal hazards; and, (3) likely to be important to the community. The attributes included the area of sandy beach, the area of foreshore reserve, the area of natural reserve, beach accessibility and presence of commercial infrastructure (Table 4.1). A cost attribute was defined in the choice experiment as a contribution to a specific coastal hazard management fund.

In the choice experiment, respondents were given a set of five 'management scenarios', describing three different management options from which they had to choose their most preferred option (see Figure 4.2). One of the options was the 'status quo', showing what changes would occur to the assets if there is no further protection. This was defined as quite severe degradation (e.g. 50 per cent loss in beach area, 75 per cent loss in the natural reserve) but would imply no additional costs. Levels of protection could be achieved by selecting a different management option, but at a personal cost to the individuals. Respondents could access information about the definition of the attributes at any point during the survey. An experimental design underpins the allocation of attribute levels in the scenarios, enabling estimation of the marginal utilities of each attribute level through random utility models such as the conditional logit (see Train, 2009).

Management scenario 1: Assuming these three options are the only ones available to you, which one would you choose? Remember to be mindful of your budget constraint.

What you will get in 10 years time:	Option 1	Option 2	Option 3 Situation in 10 years time with no management change	What you get at the moment:
 Sandy beach	50%	50%	50%	100%
 Foreshore reserve	25%	75%	50%	100%
 Natural reserve	50%	25%	25%	100%
 Beach access	Average	Good	Average	Good
 Retail, dining & club facilities	Absent	Absent	Absent	Present
Cost to you each year, for 10 years	$400	$50	$0	

Figure 4.2 Example of a management scenario for the Cottesloe Beach pilot study.

Notes
Attribute images were displayed in colour in the survey instrument; references to dollars are to AUD.

The travel cost questions were simply about the number of trips the respondents made to the beach, divided into winter and summer months, and the distance and duration of a trip from their home to the beach. The valuation questions were embedded into an online survey that also collected information about socio-demographics, and the individual's experience of Cottesloe Beach.

The primary objective of the pilot was to provide a 'proof of concept' of the survey, i.e. that a representative sample of respondents could complete the survey, and that indicative statistical results could be derived. The pilot was not intended to be an accurate reflection of results that may be derived if a full survey were implemented. The survey was launched on 22 November 2017 and data collection was completed on 23 November, via an online recruitment company that targeted the Perth metropolitan population. A total of 213 respondents entered the survey. After accounting for those who did not complete the survey, or who were excluded because quotas on age and location criteria or total number of completed responses had been met, 150 completed surveys were available for analysis.

Table 4.2 reports the estimated willingness to pay for the attributes in the survey. We identified significant values for increasing the beach area and foreshore reserve, and for the presence of commercial facilities. Given the iconic

Table 4.1 Coastal assets included in the discrete choice experiment and Cottesloe Beach pilot study

Attribute	Description	Attribute measurement	Attribute levels for Cottesloe[1] (bold text indicates the 'status quo' level if no management is undertaken)
Sandy beach	This is the area of sandy beach available for recreational use at high tide.	Percentage area	• 25% of the current beach (7,500 m²) • **50% of the current beach (15,000 m²) – i.e. the expected area in 10 years' time** • 75% of the current beach (22,500 m²) • 100% of the current beach (30,000 m²) – i.e. there is no change from today
Foreshore reserve	This is the land reserve adjacent to the sandy beach that is available for recreational use. It includes recreational facilities such as changing rooms, open grassy areas, shelter, play equipment, barbeques and picnic tables.	Percentage area	• 25% of the current foreshore reserve (3,000 m²) • **50% of the current foreshore reserve (6,000 m²) – the expected area in 10 years' time** • 75% of the current foreshore reserve (9,000 m²) • 100% of the current foreshore reserve (12,000 m²) – i.e. there is no change from today
Natural reserve	This is the area of natural reserves next to the coast, including marine ecosystems in the water near the shore and native dune vegetation.	Percentage area	• **25% of the current natural reserve (5,000 m²) – the expected area in 10 years' time** • 50% of the current natural reserve (10,000 m²) • 75% of the current natural reserve (15,000 m²) • 100% of the current natural reserve (20,000m²) – i.e. there is no change from today

Beach access	This includes the provision of pathways and steps that service the beach, as well as ramps for disability access.	Poor, average, good (e.g. number, distance between, length of or type of access points)	• Poor, with only one access point to the beach • **Average, with access points every 200 m – i.e. the expected amount in 10 years' time** • Good, with access points every 50 m – i.e. there is no change from today
Retail, dining and club facilities	This includes the provision of retail, food outlets and other public services along the foreshore reserve.	Present/absent (types of facilities affected are described)[2]	• **Absent, where current facilities deteriorate and are removed – i.e. the expected situation in 10 years' time** • Present, where current facilities are maintained – i.e. there is no change from today

Notes

1 The attribute levels were hypothetical estimates of potential changes that could occur in the quality/quantity of the coastal assets at Cottesloe in 10 years' time, due to coastal hazards or changes in management. The absolute numbers used for these levels would be tailored to the specific case study location in other applications of the survey.

2 In the case of the Cottesloe pilot study, respondents were informed that the specific facilities affected would be cafés, restaurants and surf club facilities.

Table 4.2 Willingness to pay (WTP) estimates for attributes in the Cottesloe Beach choice experiment: 2017AUD/household/year for 10 years

	WTP		*(95% confidence interval)*	
Sandy beach: per % increase in area	$2.13	★★★	0.77	3.49
Foreshore reserve: per % increase in area	$1.88	★★★	0.58	3.19
Natural reserve: per % increase in area	$1.11		−0.43	2.65
Beach access: increase from poor to average	$20.72		−76.91	118.35
Beach access: increase from poor to good	$4.50		−96.42	105.42
Facilities present	$93.73	★★	3.68	183.78

Notes
★★★, ★★ indicates significance at 99% and 95% level of confidence, respectively.

nature of Cottesloe Beach for Perth residents these results are perhaps to be expected. Changes to beach access were not significantly valued, but this again may be expected in this context: the beach is in a suburban area and access, even if restricted, would still be relatively easy to achieve, even at the lowest level of access defined in the survey.

The travel cost results presented an alternative perspective of Perth respondents' value in respect of Cottesloe Beach. Given the information on number of trips made, and the estimated travel cost of making those trips, an expected trip function can be estimated using an appropriate model for count data, and hence the value per trip inferred (Yen and Adamowicz, 1993; Parsons, 2017). The estimated value per trip is reported in Table 4.3. In the context of coastal hazard decisions, these values are only relevant if there was some hazard that would deny access to the beach in its entirety, as opposed to the marginal changes that can be valued through the choice experiment.

Conclusion

Effective decision making and investment in coastal hazard management requires an understanding of community values. Without these values, decision makers run the risk of objection towards policies and investing in suboptimal projects.

As discussed above in the section 'Measuring community preferences', a range of approaches exist to measure community values in respect of coastal assets. While decisions about coastal hazard investment are commonly

Table 4.3 Willingness to pay (WTP) estimates from seasonal travel cost models for Cottesloe Beach

	Summer model *(95% confidence interval)*	*Winter model* *(95% confidence interval)*
WTP per trip (2017 AUD)	$2.08 (0.74–3.42)	$1.73 (0.87–2.56)

resource-limited, non-market valuation approaches that can quantify community values in monetary-equivalent terms are particularly useful to enable optimal prioritisation of different investment options.

The pilot study presented illustrates how non-market valuation approaches can successfully be used to measure values for coastal assets that may be at risk of coastal hazards (Rogers and Burton, 2018). What the study indicated was that it is possible to develop a stated and revealed preference survey, framed so that it is adaptable to multiple coastal locations in Western Australia. This makes the approach accessible for decision makers to utilise in a consistent and comparable manner across different locations and hazard management projects. There are ongoing discussions with Western Australia's Department of Planning, Lands and Heritage and local town councils (who are responsible for the implementation of the CHRMAP process) about extending the study from a pilot to a case study for evaluating alternative management outcomes.

Growing populations relying on coastal assets, and increased environmental threats to those assets, mean that efficient investment in coastal hazard management will become an increasingly important issue for governments. Part of the response to this challenge will be the development and use of economic tools, such as those described here, that can inform decision makers about which management outcomes will improve the welfare of the communities they are entrusted to protect.

Acknowledgements

The authors would like to acknowledge the Western Australian Department of Planning, Lands and Heritage and the Western Australian Planning Commission for their funding and support for this research, as well as the Local Government Associations who participated in the development of the survey instruments and associated guidance material. Thanks also to the editors of this book, Erika Techera and Gundula Winter, as well as Sam Bishopp and Vivienne Panizza for their comments which have improved this chapter.

Note

1 Benefit transfer is often used when the time and resources are not available to conduct an original non-market valuation study. It is a process where a value estimated by an existing study is applied to a new, similar decision or policy context (see Johnston *et al.*, 2015 for more information).

References

Alves, B., Rigall-I-Torrent, R., Ballester, R., Benavente, J. and Ferreira, O. (2015). 'Coastal erosion perception and willingness to pay for beach management (Cadiz, Spain)'. *Journal of Coastal Conservation*, vol 19, pp. 269–280.
Bateman, I.J., Carson, R.T., Day, D., Hanemann, M., Hanley, N., Hett, T., Jones-Lee, M., Loomes, G., Mourato, S., Özdemiroglu, E., Pearce, D.W., Sugden, R.

and Swanson, J. (2002). *Economic Valuation with Stated Preference Techniques: A Manual*. Cheltenham, UK: Edward Elgar.

Bennett, J. and Blamey, R. (eds.) (2001). *The Choice Modelling Approach to Environmental Valuation*. Cheltenham, UK: Edward Elgar.

Berz, G., Kron, W., Loster, T., Rauch, E., Schimetschek, J., Schmieder, J., Siebert, A., Smolk, A. and Wirtz, A. (2001). 'World map of natural hazards: A global view of the distribution and intensity of significant exposures'. *Natural Hazards*, vol 23, no 2–3, pp. 443–465.

Bishop, R.C., Boyle, K.J., Carson, R.T., Chapman, D., Hanemann, W.M., Kanninen, B., Kopp, R.J., Krosnick, J.A., List, J., Meade, N., Paterson, R., Presser, S., Smith, V.K., Tourangeau, R., Welsh, M., Wooldridge, J.M., DeBell, M., Donovan, C., Konopka, M. and Scherer, N. (2017). 'Putting a value on injuries to natural assets: The BP oil spill'. *Science*, vol 356, no 6335, pp. 253–254.

Carson, R.T. (2011). *Contingent Valuation: A Comprehensive Bibliography and History*. Cheltenham, UK: Edward Elgar.

Carson, R.T., Mitchel, R.C., Hanemann, M., Kopp, R.J., Presser, S. and Rudd, P.A. (2003). 'Contingent valuation and lost passive-use: Damages from the Exxon Valdez oil spill'. *Environmental and Resource Economics*, vol 25, pp. 257–286.

Christie, M., Fazey, I., Cooper, R., Hyde, R. and Kenter, J.O. (2012). 'An evaluation of monetary and non-monetary techniques for assessing the importance of biodiversity and ecosystem services to people in countries with developing economies'. *Ecological Economics*, vol 83, pp. 67–78.

Department of Climate Change (2009). 'Climate change risks to Australia's coast: A first pass national assessment'. Department of Climate Change, Australian Government, Canberra.

Dribek, A. and Voltaire, L. (2017). 'Contingent valuation analysis of willingness to pay for beach erosion control through the stabiplage technique: A study in Djerba (Tunisia)'. *Marine Policy*, vol 86, pp. 17–23.

ERG (2016). 'Hurricane Sandy and the value of trade-offs in coastal protection and restoration'. Report prepared for NOAA Office for Coastal Management, Silver Spring, MD.

Hanley, N. and Barbier, E.B. (2009). *Pricing Nature: Cost–Benefit Analysis and Environmental Policy*. Cheltenham, UK: Edward Elgar.

Huang, J-C., Poor, P.J. and Zhao, M.Q. (2007). 'Economic valuation of beach erosion control'. *Marine Resource Economics*, vol 22, pp. 221–238.

Johnston, R.J., Rolfe, J., Rosenberger, R.S. and Brouwer, R. (eds.) (2015). *Benefit Transfer of Environmental and Resource Values: A Guide for Researchers and Practitioners*. Dordrecht: Springer.

Johnston, R.J., Boyle, K.J., Adamowicz, W., Bennett, J., Brouwer, R., Cameron, T.A., Hanemann, W.M., Hanley, N., Ryan, M., Scarpa, R., Tourangeau, R. and Vossler, C.A. (2017). 'Contemporary guidance for stated preference studies'. *Journal of the Association of Environmental and Resource Economists*, vol 4, no 2, pp. 319–405.

Kirkpatrick, S. (2011). 'The economic value of natural and built coastal assets. Part 1: Natural coastal assets'. Discussion paper, Australian Climate Change Adaptation Research Network for Settlements and Infrastructure.

Kirkpatrick, S. (2012). 'The economic value of natural and built coastal assets. Part 2: Built coastal assets'. Discussion paper, Australian Climate Change Adaptation Research Network for Settlements and Infrastructure.

Kontogianni, A., Damigos, D., Tourkolias, C., Vousdoukas, M., Velegrakis, A., Zanou, B. and Skourtos, M. (2014). 'Eliciting beach users' willingness to pay for protecting European beaches from beachrock processes'. *Ocean and Coastal Management*, vol 98, pp. 167–175.

Londoño, L.M. and Johnston, R.J. (2012). 'Enhancing the reliability of benefit transfer over heterogeneous sites: A meta-analysis of international coral reef values'. *Ecological Economics*, vol 78, pp. 80–89.

Matthews, Y., Scarpa, R. and Marsh, D. (2017). 'Using virtual environments to improve the realism of choice experiments: A case study about coastal erosion management'. *Journal of Environmental Economics and Management*, vol 81, pp. 193–208.

Melillo, J.M., Richmond, T.C. and Yohe, G.W. (eds.) (2014). 'Climate change impacts in the United States: The third national climate assessment'. US Global Change Research Program, Washington, DC, doi:10.7930/J0Z31WJ2.

Neumann, B., Vafeidis, A.T., Zimmermann, J. and Nicholls, R.J. (2015). 'Future coastal population growth and exposure to sea-level rise and coastal flooding – A global assessment'. *PLoS ONE*, vol 10, p. e0118571, doi:10.1371/journal.pone.0118571.

Pandit, R., Subroy, V., Garnett, S.T., Zander, K.K. and Pannell, D. (2015). 'A review of non-market valuation studies of threatened species and ecological communities'. Report to the National Environmental Science Programme, Department of the Environment, Canberra.

Parsons, G.R. (2017). 'Travel cost models', in P. Champ, K. Boyle and T. Brown (eds.), *A Primer on Non-Market Valuation*. Dordrecht: Springer.

Perman, R., Ma, Y., McGilvray, J. and Common, M. (1999). *Environmental and Natural Resource Economics*. London: Addison Wesley Longman.

Peters, H. and Hawkins, J.P. (2009). 'Access to marine parks: A comparative study in willingness to pay'. *Ocean & Coastal Management*, vol 52, pp. 219–228.

Policy Research Corporation (2009). 'The economics of climate change adaptation in EU coastal areas'. Study carried out on behalf of the Directorate-General for Maritime Affairs and Fisheries, European Commission, Brussels.

Polomé, P. (2002). 'Extracting a benefit transfer function from CV studies: Final report'. DELOS WP 4.1, EU Fifth Framework Programme, Energy, Environment and Sustainable Development. Available at: www.delos.unibo.it/Docs/Deliverables/D11%20Extracting%20a%20Benefit%20transfer%20Function%20from%20CV%20studies.pdf.

Rogers, A. and Burton, B. (2018). 'Non-market valuation instruments for measuring community values affected by coastal hazards and their management'. Report prepared for the Western Australian Department of Planning, Lands and Heritage, The University of Western Australia, Crawley.

Rogers, A., Kragt, M., Gibson, F., Burton, M., Petersen, E. and Pannell, D. (2015). 'Non-market valuation: Usage and impacts in environmental policy and management in Australia'. *The Australian Journal of Agricultural and Resource Economics*, vol 59, no 1, pp. 1–15, doi:10.1111/1467-8489.12031.

Torres, C. and Hanley, N. (2016). 'Economic valuation of coastal and marine ecosystem services in the 21st century: An overview from a management perspective'. Stirling Economics Discussion Paper 2016-01, University of Stirling, Scotland.

Train, K.E. (2009). *Discrete Choice Methods with Simulation* (2nd edn.). New York: Cambridge University Press.

Western Australian Government (2017). 'WA coastal zone strategy'. Western Australian Planning Commission, Perth.

Western Australian Planning Commission (2014). 'Coastal hazard risk management and adaptation planning guidelines'. Department of Planning, Western Australian Planning Commission, Perth.

Windle, J. and Rolfe, J. (2014). 'Estimating the nonmarket economic benefits of beach resource management in southeast Queensland, Australia'. *Australasian Journal of Environmental Management*, vol 21, no 1, pp. 65–82.

Wong, P.P., Losada, I.J., Gattuso, J.-P., Hinkel, J., Khattabi, A., McInnes, K.L., Saito, Y. and Sallenger, A. (2014). 'Coastal systems and low-lying areas', in *Climate Change 2014: Impacts, Adaptation, and Vulnerability. Part A: Global and Sectoral Aspects. Contribution of Working Group II to the Fifth Assessment Report of the Intergovernmental Panel on Climate Change* [C.B. Field, V.R. Barros, D.J. Dokken, K.J. Mach, M.D. Mastrandrea, T.E. Bilir, M. Chatterjee, K.L. Ebi, Y.O. Estrada, R.C. Genova, B. Girma, E.S. Kissel, A.N. Levy, S. MacCracken, P.R. Mastrandrea and L.L.White (eds.)]. Cambridge and New York: Cambridge University Press, pp. 361–409.

Yen, S.T. and Adamowicz, W.L. (1993). 'Statistical properties of welfare measures from count data models of recreation demand'. *Review of Agricultural Economics*, vol 15, no 2, pp. 203–215.

5 Nature-based solutions to mitigate extreme coastal impacts

Gundula Winter, Karin R. Bryan and Marco Ghisalberti

Introduction

Coasts are dynamic zones at the interface between the land and the ocean and are constantly evolving due to the impact of waves, wind and tides. During marine extreme events such as coastal storms and tsunamis, these changes can take place over a short timespan from a few minutes to several days. The coastal zone (up to 100 km inland) also supports a population that is three times denser than the global average (Small and Nicholls, 2003), and 13 per cent of the global urban population now lives less than 10 m above mean sea level (McGranahan *et al.*, 2007). Many of these communities depend upon natural habitats found in the coastal zone for food, fuel and livelihood. Yet the combination of dynamic natural processes and a high population density contributes to the high risk that characterises the coastal zone. The vulnerability of coastal settlements has been highlighted by a series of historic extreme events including the North Sea storm surge in 1953, the Boxing Day tsunami in the Indian Ocean in 2004 and Hurricane Katrina in the US in 2005 (Kathiresan and Rajendran, 2005). As extreme sea levels are expected to heighten with mean global sea level rise (Wahl *et al.*, 2017), storms are expected to intensify (Young *et al.*, 2011) and coastal populations are likely to increase (Small and Nicholls, 2003), so coastal risk is on the rise.

Coastal protection will need to be reinforced in many places to safeguard populations and assets within the coastal zone from the impact of future extreme events. Traditional coastal engineering approaches to protect the coast from flooding and erosion rely on hard structures such as dikes, sea walls and revetments (Borsje *et al.*, 2011). These solutions are designed for a limited lifespan and adapt poorly to changing extreme conditions. Their construction and maintenance are often costly and negatively affect recreation, water quality and fisheries (Arkema *et al.*, 2013). In estuaries, for example, engineered shorelines can even increase flood levels (Temmerman *et al.*, 2013). More recently, attention has shifted to nature-based solutions for coastal protection (Bouma *et al.*, 2014). Natural systems, such as mangroves, seagrass, kelp, saltmarshes and reefs, reduce wave energy and help to stabilise or even grow the coastline. This is due to the capacity of vegetation to trap and stabilise sediment (de Boer, 2007) as well as to produce new sediment through organic input, which allows these systems to adapt to rising sea levels (McKee *et al.*, 2007). In addition to affecting elevation, natural systems can adapt their growth forms to changing environmental conditions and, if given enough time, can even regenerate after extreme events (Paling *et al.*, 2008). Nature-based solutions often come at no or lower cost than structural solutions (Barbier, 2013) and different species are available to suit climates ranging from the tropics to temperate zones. It is therefore critically important that those tasked with planning, approving and implementing coastal protection measures are fully informed of the benefits and shortcomings of both engineered and nature-based options.

The concept of nature-based solutions is not new and has been widely explored all over the world. Salt marshes, for example, have been used for dyke protection and land reclamation for centuries (Gedan *et al.*, 2009), and dunes have been planted to protect beaches from erosion. Nature-based coastal protection can be an integral part of ecosystem-based management and positively contribute to multiple other human development sectors such as food security, sustainable water management and livelihood diversification (Jones *et al.*, 2012; Morris *et al.*, 2018). Currently, widely differing terminology is used to describe similar concepts. In the Netherlands, the 'Building with Nature' programme is a broader approach to coastal protection, which encapsulates all 'soft' measures including coastal vegetation but also sediment supply to the coast (van Slobbe *et al.*, 2013). In the US, 'Living Shorelines' refers to a narrower definition of using natural organisms to protect and add value to the coast (e.g. O'Donnell, 2016). Relatively little literature seeks to communicate these concepts to governments, private industry and coastal communities which must make decisions to protect coastal areas for the future. This chapter seeks to demonstrate the coastal protection value of natural habitats.

Background

Coastal hazards

For the purpose of this chapter we define marine extreme events as storms, large swell-wave events and tsunamis. Storms can be distinguished as tropical cyclones or mid-latitude storms. Tropical cyclones typically have a large pressure gradient but a small footprint. Their impact is often catastrophic but confined to a small stretch of coast. In contrast, mid-latitude storms associated with weather systems in the westerlies zone typically have a much larger footprint but lower peak wind velocities.

The low air pressure within storm systems reduces the pressure that acts on the sea surface, which causes the sea level to rise and the so-called storm surge. Additionally, an onshore directed wind can contribute further to the height of the storm surge. Wind set-up is inversely proportional to the water depth and is thus most important on large and shallow continental shelves. Storm surge levels are in addition to astronomic tide levels and the most critical conditions arise when high tide and peak storm surge levels coincide.

When wind blows across a body of water, it generates waves whose height depends on wind strength, duration and the size of the area the wind is acting on (wind fetch). As waves approach a coastline and enter shallow water they steepen and heighten until they eventually break. When waves break they create a force, which causes a set-up of the water level along the coastline in addition to the storm surge described above. Broken waves continue to propagate towards the shore as white-water bores, eventually moving up the shore face as wave run-up. Wave run-up height depends on the shoreline slope and contributes to overtopping of coastal protection such as dunes and dykes and subsequent coastal flooding. Additionally, wave components with different wave periods interact to create a rhythmic set-up and set-down of the water level of the order of minutes, which is commonly referred to as surf beat (Tucker, 1950). This surf beat can add to the instantaneous water level and increase maximum wave run-up and surge levels at the shoreline.

During storms, large waves erode sediment from the beach face and dunes and near bed currents transport this sediment offshore, where it is deposited as a sand bar. During calmer wave conditions, waves push this sediment back towards the beach (Wright *et al.*, 1984). The shape of the beach is thus always the product of the history of wave conditions (Ludka *et al.*, 2015). Waves that approach the shoreline at an angle generate alongshore currents that transport sediment along the coast, which can lead to localised erosion and accretion patterns.

Tsunamis are long waves in the ocean that are either generated by an earthquake or a subsea landslide. Tsunami waves propagate at a speed dependent on the water depth. Hence, in the deep ocean, they propagate at very fast speeds (~200 km/hr) while their wave amplitude is very small O(cm). As they enter shallow water, tsunami waves rapidly slow down and simultaneously increase in height.

Canopy flow

The impact of benthic ecosystems such as mangroves, seagrass, kelp, salt-marshes and reefs on flows and waves in coastal systems can be profound. Clearly, the biological functions of these systems are distinct. Furthermore, there is a highly variable morphology, with systems ranging from rigid (e.g. mangroves, most reefs) to very flexible (e.g. seagrass and kelp), and from submerged in the water to emerging fully from the water surface. Despite these differences, all of these ecosystems have the physical effect of exerting drag, and there are thus commonalities across all systems in terms of the impact they have on the flow. In doing so, they can greatly influence the structure of mean currents and surface waves, rendering both completely distinct from those over bare beds (Ghisalberti, 2009). The capacity of benthic canopies to attenuate flows, flooding and coastal erosion is therefore typically in proportion to their ability to exert drag. Thus, these services tend to increase with the density of the canopy (often expressed as a frontal area per unit volume) and its height relative to the water depth (variable during extreme events due to storm surge and wave set-up), and decrease with canopy flexibility (see, for example, McDonald *et al.*, 2006; Koch *et al.*, 2009).

In both current- and wave-dominated systems, the canopy drag reduces the near-bed flow velocity relative to that over a bare bed. This reduction can be by an order of magnitude or more (Gambi *et al.*, 1990; Lowe *et al.*, 2005b). This, in turn, greatly diminishes the stress exerted by the flow on the sediment bed (Hansen and Reidenbach, 2012; Pomeroy *et al.*, 2017). Accordingly, these systems are typically regions of reduced sediment erosion and enhanced sediment deposition (Montakhab *et al.*, 2012; Nepf, 2012). While there are exceptions to this rule (e.g. van Katwijk *et al.*, 2010; Lawson *et al.*, 2012), the majority of benthic ecosystems have the potential to trap large volumes of sediment in coastal systems.

Benthic ecosystems can also influence the properties of surface waves that propagate across them. Notably, canopy drag can cause strong decreases in wave height, resulting in much smaller waves at the shoreline. This wave attenuation increases with the density of the canopy and its height relative to the water depth (Kobayashi *et al.*, 1993; Hansen and Reidenbach, 2013; Morris *et al.*, 2018). Significant wave attenuation is typically seen over distances of O (10–100 m) (Henderson *et al.*, 2017). The canopies preferentially attenuate shorter waves (Lowe *et al.*, 2007), increasing the mean period of the wave field. In cases where the canopy is very dense and takes up a significant fraction of the water depth (particularly in reef systems), the effective shallowing can induce wave breaking (Hearn, 1999). Through these mechanisms, benthic canopies significantly diminish the wave heights to which the shoreline is exposed.

Through the drag they exert, emergent canopies in coastal wetlands can be particularly effective at counteracting coastal flooding. This occurs largely due to reductions in storm surge and wave run-up (Spalding *et al.*, 2014; Spencer *et al.*, 2014). Submerged canopies in deeper water are less effective in this regard, although the wave attenuation created by dense, tall canopies has a direct follow-on effect in reducing wave run-up (Løvås and Tørum, 2001; John *et al.*, 2016).

The overall hydrodynamic influence of these ecosystems, in reducing wave height, diminishing coastal flooding and trapping sediment, lends itself to such systems being important contributors to coastal defence. Overall, there is strong evidence for long-term enhanced coastal protection due to natural (Barbier *et al.*, 2011; Spalding *et al.*, 2014; Guannel *et al.*, 2016) and restored (Morris *et al.*, 2018) benthic ecosystems. However, most of the work in this area focuses on persistent defence over extended periods of time. The coastal protection provided during extreme events (such as storms, cyclones and tsunamis) is not so well established (Morris *et al.*, 2018). In the following section, common benthic ecosystems are presented with a focus on their suitability to protect the coast from flooding and erosion during extreme events.

Ecosystems in nature-based coastal defences

Mangroves

Mangroves grow along sub-tropical and tropical shorelines, existing between mean sea level and high tide. They are currently declining globally (Thomas *et al.*, 2017), due to changes in land use (e.g. into aquaculture or for coastal development), harvesting (Kathiresan and Bingham, 2001), sea level fluctuations (Lovelock *et al.*, 2017) and, in rare cases, removal as an invasive weed (e.g. in New Zealand (Horstman *et al.*, 2018)). Mangrove morphology and growth characteristics vary vastly with species and environment. For example, *Avicennia* and *Sonneratia* often grow on the seaward fringe, with single trunks and pencil root-like structures protruding from lateral cable roots within the seabed, whereas the very complex *Rhizophera* have dense umbrella-like prop roots emanating from a central trunk. Their roots and trunks are rigid (although in some species, e.g. *Avicennia*, roots can vibrate with wave action), whereas the canopies are flexible (although often not submerged).

Mangroves have been shown to reduce energy from waves, storm surges and tsunamis. Although dissipation of wave energy has been demonstrated in many recent studies (e.g. Massel *et al.*, 1999; Mazda *et al.*, 2006; Quartel *et al.*, 2007; Bao, 2011; Horstman *et al.*, 2014), cases demonstrating reduction of storm waves and storm surge in the field are far less common. Dissipation rates are often inferred from laboratory measurements (e.g. Kobayashi *et al.*, 1993), or numerical modelling studies (Chang *et al.*, 2017; Parvathy *et al.*, 2017) that have been calibrated using observations. Field measurements indicate that mangroves can reduce wave energy by approximately 50 per cent per 100 m (summarised in McIvor *et al.*, 2012), but measurements vary between sites because of differences in vegetation characteristics, bed slope and wave characteristics (Vo-Luong and Massel, 2008; Bao, 2011). Most studies show dissipation is greatest near the fringe (e.g. Norris *et al.*, 2017).

The few observations that show a reduction in storm surge by mangroves show conflicting estimates. For example, Krauss *et al.* (2009) show reduction in the order of 10 cm per kilometre, whereas Zhang *et al.* (2012) found much larger values of up to 50 cm per kilometre. Channelisation, patchiness of the vegetation and the existence of small lakes and ponds within the forest likely account for differences. Both authors emphasise that mangroves can increase surge amplitudes by causing shoaling as the surge wave propagates into the vegetated zone. Recent studies have also shown that the contribution of wave set-up during storm events could add a significant amount to the storm surge (van Rooijen *et al.*, 2016), where the set-up is induced by the gradient in wave height caused by vegetation-induced dissipation.

Similarly, *in situ* hydrodynamic measurements of tsunamis are rare, and most studies on the ability of mangroves to attenuate tsunamis waves are inferred by correlating zones of damage to the extent of mangroves (e.g. Kathiresan and Rajendran, 2005; Baird *et al.*, 2009; Das and Vincent, 2009). However, some estimates of flow velocity during the 2004 Boxing Day tsunami were extracted from survivor videos (Fritz *et al.*, 2006). In addition to dissipating energy, forests can stop drifting ships and flotsam, trap people before they are swept out to sea and trap sediment to create an additional barrier (Harada and Imamura, 2005). However, recent reports suggest that tsunami-protection characteristics have been overstated (Baird *et al.*, 2009). Laboratory experiments are often used to study tsunami dissipation (Strusinska-Correia *et al.*, 2013; Tsai *et al.*, 2017). The controlled environments allow the effect of slope-induced breaking to be separated from vegetation-induced breaking, and show that often the dissipation occurs in a surfzone seaward of the vegetation (Strusinska-Correia *et al.*, 2013). In a very large tsunami such as the Great East Japan Tsunami, not even traditional hard solutions provided adequate protection, and coastal protection provided by forests was negligible. Although mangroves are predicted to recover quickly due to below ground nutrient and organic stocks (Alongi, 2008), binding of sediments by root mass makes them more resistant to erosion during events, facilitating recovery. However, re-establishing and conserving mangrove forests in urban areas can pose a public health risk due to the increased incidence of mosquito-borne diseases (Claflin and Webb, 2017).

Seagrass

Seagrasses cover 0.1–0.2 per cent of the global ocean (Duarte, 2002), and provide a range of ecosystem services. These services include nutrient uptake, high rates of productivity resulting in oxygen provision and carbon sequestration, enhanced biodiversity and coastal protection (Duarte, 2002). As a result of these ecosystem services, the value of seagrass meadows was estimated (two decades ago) at US$19,000 per hectare per year (Costanza *et al.*, 1997), exceeding the estimated values of coral reefs, mangroves and tropical forests. As photosynthetic organisms, seagrasses face increasing pressures from high-turbidity waters, such as those due to dredging plumes, sediment pulses from storm events and sediment-laden river discharges (Orth *et al.*, 2006).

Seagrasses exhibit 'ecosystem engineering' behaviour in their capacity to enhance sediment settling and retention. In particular, the complex network of subsurface rhizomes increases the critical shear stress required to remobilise sediment in seagrass meadows (Christianen *et al.*, 2013). Thus, seagrasses maximise their sediment retention potential by diminishing the actual stress on the sediment bed and increasing the threshold stress required to mobilise it.

Seagrasses have several characteristics that potentially limit their capacity to provide coastal protection in extreme conditions. First, seagrasses (across all species) exhibit significant flexibility. This means that seagrass canopies tend to be relatively pronated in strong flows, exerting less drag and potentially limiting the coastal protection services they can provide during extreme events. Second, in temperate regions, seagrasses exhibit a pronounced senescence during the colder months (Hansen and Reidenbach, 2013). In many coastal regions, therefore, the above-ground biomass of seagrass meadows is not in-phase with the likelihood of storms; that is, in terms of coastal protection, the maximum seagrass coverage is not present when it is most needed (Koch *et al.*, 2009). Third, seagrass meadows are themselves extremely vulnerable to extreme events. The strong flows and large waves that arise in extreme conditions leave seagrasses susceptible to physical uprooting and blade breaking (Fonseca *et al.*, 2007). Furthermore, there is a complex feedback loop between seagrass biomass, light intensity and suspended sediment levels (Adams *et al.*, 2016); the sediment resuspension that occurs during extreme events can thus have significant knock-on effects on seagrass stability (Kilminster *et al.*, 2015).

Despite this behaviour, it has been shown that seagrass meadows can provide critical coastal protection during extreme events. For example, seagrass meadows were shown to have a strong buffering effect against the 2004 Boxing Day tsunami in Aceh and Southern Thailand. Across 623 sites impacted by the tsunami, the flooded area was significantly lowered in the presence of seagrass beds (Chatenoux and Peduzzi, 2007). The consistent buffering effect of seagrass beds against the tsunami contrasts with the impact of coral reefs, which were often seen to accentuate flooding (Cochard *et al.*, 2008). The modelling of Vu *et al.* (2017) suggests that for decadal to centennial storms off the French coast in the Mediterranean, the drag exerted by *Posidonia oceanica* meadows decreased the significant wave height by up to 25 per cent and rates of net sediment transport by up to 60 per cent. This is consistent with the observed attenuation effect of *Posidonia oceanica* meadows on currents and waves and the promotion of sediment stability during a storm off the Spanish coast (Granata *et al.*, 2001).

To our knowledge, there have not been any studies into the coastal protection services afforded by restored seagrass beds. This is a critical gap, as the costs and benefits of natural system restoration for coastal defence (relative to the construction of engineered systems) remain difficult to quantify. There remains significant uncertainty about threshold values of scale and density that are required for long-term seagrass restoration success, particularly in the capacity of a restored meadow to withstand periods of high flow during the establishment period (Katwijk *et al.*, 2016).

Kelp

Kelp are a characteristic of temperate and arctic waters (Krumhansl *et al.*, 2016), and tend to exist near zones of upwelling where the nutrient supply is adequate to fuel these large growth forms. Although there are no consistent trends in global kelp abundance (Krumhansl *et al.*, 2016), there are many studies indicating that kelp forests are under stress (e.g. Connell *et al.*, 2008; Filbee-Dexter *et al.*, 2016). Very few studies show the ability of kelp to significantly reduce energy. Although there are several studies showing that kelp is often associated with reduced tidal and shelf currents (Jackson and Winant, 1983; Jackson, 1984, 1997; Rosman *et al.*, 2007), and early experimental work showed attenuation of waves (Dubi and Tørum, 1995), the work of Elwany *et al.* (1995) shows no effect of kelp on waves in field observations. Clearly the species of kelp (*Laminaria hyperborea* in the former study and *Macrocystis pyrifera* in the latter) and differences in stem rigidity play a role. Kelp consists of flexible stems (stipes) onto which multiple flexible leaves (fronds) are attached. Thus, the kelp bends in response to hydrodynamic forcing, rather than applying the resistance needed to dissipate energy (such as in rigid vegetation). Due to limited research on kelp's performance during extreme events, its potential to increase coastal resilience is unknown.

Saltmarshes

Saltmarshes occur in middle to high latitudes where they occupy the upper areas of intertidal mudflats. Saltmarshes consist of herbs, grasses and shrub and prefer sheltered areas, for example, in estuaries or in the lee of barrier islands (Mcowen *et al.*, 2017). Saltmarshes trap and stabilise sediment, which enhances plant growth creating a positive feedback loop (van de Koppel *et al.*, 2005). Land reclamation, tidal restriction, pollution and invasive species disturb this process and have contributed to a global decline in saltmarshes (Gedan *et al.*, 2009). The ability to accrete land vertically allows saltmarshes to adjust to rising sea levels by migrating upward and landward. However, where the landward migration is impeded by coastal structures, saltmarshes are lost to 'coastal squeeze' (Wolters *et al.*, 2005). The decline in saltmarshes leaves coasts vulnerable to erosion; for example, in the Mexican Gulf, saltmarsh resilience was compromised following the BP 'Deepsea Horizon' oil spill as marsh edges eroded and platform heights were reduced, which hampered their recovery (Silliman *et al.*, 2012).

Saltmarshes reduce wave heights by up to 72 per cent (Narayan *et al.*, 2016), depending on inundation height, standing biomass, vegetation density, marsh size and plant stiffness (Shepard *et al.*, 2011; Bouma *et al.*, 2014). During storm surges, water levels are elevated and saltmarshes are typically submerged, but they still attenuate wave energy through frictional dissipation instead of breaking (Möller *et al.*, 2014). Under the waves, submerged flexible canopies flatten and, thus, protect the bed from erosion. This effect also makes flexible plant species highly resilient to storm wave damage (Spencer *et al.*, 2015). Stiffer canopies, in contrast, attenuate more wave energy but are also prone to folding and breakage after which their wave attenuation capacity is reduced (Rupprecht *et al.*, 2017). Still, the plant detritus provides added roughness, which dissipates wave energy and protects the bed from erosion. This keeps the root system intact and enables plant recovery in the next growing season (Fagherazzi, 2014). Although not sufficiently studied, saltmarshes are also likely to reduce flood levels because they provide accommodation space and drainage (Shepard *et al.*, 2011).

Reefs

Coral reefs are most abundant in the Indo-Pacific (Spalding *et al.*, 2001) and typically rise steeply from deep water to shallow platforms that either fringe the shoreline or are separated from the shoreline by a lagoon of variable size (Falter *et al.*, 2013). The surface of the reef structures is rough (Rosman and Hench, 2011), which reduces incoming wave heights not only by breaking but also through frictional losses over the shallow reef structures (Lowe *et al.*, 2005a). The wave sheltering effect of reefs typically promotes the accretion of sediment and the formation of shoreline cusps (Sanderson and Eliot, 1996). In addition, reefs produce carbonate sediment and thus contribute to island and beach growth (Perry *et al.*, 2008). The roughness of the reef, which contributes largely to the reef's ability to dissipate energy, is strongly dependent on the health and species assemblage of reefs (Harris *et al.*, 2018). However, local human pressures and, more importantly, rising sea surface temperatures are causing corals to die off (Hughes *et al.*, 2017) (see also Chapter 8). The complex branching corals, which cause high friction, are particularly vulnerable to the changing environmental conditions (Renema *et al.*, 2016).

While reefs are very effective in wave attenuation, they cannot alleviate surge levels. In contrast, the rough surface even increases the wave set-up over the reef (Buckley *et al.*, 2016). During extreme events, the elevated water level on the reef due to storm surge and wave set-up allows larger waves to propagate across the reef (Kench and Brander, 2006). The specific form of reefs with a steep drop-off at the offshore side and a limited distance to the shoreline can cause the surf beat motion to resonate during storm conditions. Resonant conditions amplify the surf beat amplitude at the shoreline and reportedly caused severe flooding during Tropical Cyclone Man-Yi at Guam (Péquignet *et al.*, 2009) and Tropical Cyclone Haiyan at the Philippines (Roeber and Bricker, 2015).

Oyster reefs provide a similar capacity to dampen waves and stabilise sediment (Piazza *et al.*, 2005) and can thus replace coastal revetments. However, due to food competition, mussels and oysters can be cultivated only on a limited scale (Hertweck and Liebezeit, 2002).

Conclusions

The efficacy of natural systems to dampen wave energy and mitigate coastal impacts is well known and the economic value of this coastal protection service has been quantified (e.g. Koch *et al.*, 2009). In fact, in many instances naturally existent habitats already provide a large coastal protection benefit as they mitigate coastal flooding and erosion. In the US alone, the coastal protection service provided by saltmarshes is estimated at 23 billion US dollars (USD) (Costanza *et al.*, 2008). And in the Maldives, coral reefs contribute coastal protection benefits worth 1.6 to 2.7 billion USD, while reef conservation would cost only 47 million USD per year (Jones *et al.*, 2012).

Yet, nature-based solutions are not widely implemented when designing new coastal protection solutions because the frameworks are lacking that could translate scientific knowledge into management and design guidelines. The problem is that nature-based solutions for coastal protection are desirable because their dynamic behaviour makes them adaptive to changing environmental conditions, but also less predictable with respect to their long-term reliability (Bouma *et al.*, 2014). Furthermore, the suitability for mitigating flooding and erosion differs among species. Some species can protect against smaller events, but fail in the extremes associated with storms and tsunamis. Growth pattern, density and patchiness can also limit effectiveness. These shortcomings make it challenging for governments and other decision-makers to favour nature-based solutions over engineered ones. To address these issues, the performance and resilience of all species, particularly during extreme conditions, needs to be monitored (see Chapter 6) and modelling capabilities need to be improved (Borsje *et al.*, 2011). Tipping points need to be identified below which ecosystems can provide risk reduction services taking into account the impact of climate change, invasive species and human pressures (Jones *et al.*, 2012) as well as habitat alterations due to extreme events (Howes *et al.*, 2010; Spalding *et al.*, 2014). Recovery of natural systems after extreme events can be long and compromised by poor health, but can be enhanced by interventions such as replanting (Borja *et al.*, 2010). Uncertain performance under extreme conditions and long recovery means that nature-based solutions may not always be adequate depending on the local geomorphology, oceanographic boundary conditions, habitat type and size and local risk levels. Additionally, ecosystem-based coastal defence options require space, which may necessitate managed realignment of infrastructure along developed shorelines (Wolters *et al.*, 2005). This may lead to socio-economic tensions where communities are expanding and seeking to develop the coastal zone for housing, tourism or other purposes. Where space requirements or uncertainty levels are unacceptable, engineering solutions may need to be considered, preferably in combination with nature-based approaches in so-called hybrid solutions (Spalding *et al.*, 2014).

Weighing up the costs and benefits of all options always needs to be attempted at a local scale, modelling the physical hazards and ecology, but also taking into account other services that these ecosystems provide. Many coastal habitats can contribute to fisheries as they provide important birthing and nursery grounds; mangroves, seagrasses and saltmarshes can sequester a large amount of carbon depending on their species and underground root stores (Duarte, 2002). Indeed, recent attention has been given to the value of coastal vegetation as carbon sinks leading to 'blue carbon' initiatives to mitigate the impacts of greenhouse gas emissions. Given the multiple values associated with mangroves, seagrasses, kelp, saltmarshes and reef environments, the coastal protection options available need to be evaluated in collaboration with the social sciences. Only multi-sectoral and multi-disciplinary investigations can address the challenges associated with competing uses for the natural resources and spaces in the coastal zone.

This chapter demonstrates the potential value of natural environments for coastal protection, which adds to the imperative to protect stocks and sites so they also remain intact for the other ecosystem services they provide. Where ecosystems have been lost, targeted restoration can be an option to re-establish coastal protection and other ecosystem services. Multi-disciplinary research, including social science research, is also needed to focus restoration efforts on locations where hazards can be effectively reduced and where people can gain the most (Arkema *et al.*, 2013). Finally, clear communication on the ecosystem services that different species can provide and under which conditions (including wave attenuation, flood level reduction or sediment retention) is important to create adequate public perceptions on the benefits and shortcomings of nature-based solutions. This research and the communication thereof are of critical importance to inform government decision-makers tasked with coastal protection in the face of a changing climate and growing coastal populations.

References

Adams, M. P., Hovey, R. K., Hipsey, M. R., Bruce, L. C., Ghisalberti, M., Lowe, R. J., Gruber, R. K., Ruiz-Montoya, L., Maxwell, P. S. and Callaghan, D. P. (2016). 'Feedback between sediment and light for seagrass: Where is it important?'. *Limnology and Oceanography*, vol 61, no 6, pp. 1937–1955.

Alongi, D. M. (2008). 'Mangrove forests: Resilience, protection from tsunamis, and responses to global climate change'. *Estuarine, Coastal and Shelf Science*, vol 76, no 1, pp. 1–13. doi:10.1016/j.ecss.2007.08.024.

Arkema, K. K., Guannel, G., Verutes, G., Wood, S. A., Guerry, A., Ruckelshaus, M., Kareiva, P., Lacayo, M. and Silver, J. M. (2013). 'Coastal habitats shield people and property from sea-level rise and storms'. *Nature Climate Change*, vol 3, no 10, pp. 913–918. doi:10.1038/nclimate1944.

Baird, A. H., Bhalla, R. S., Kerr, A. M., Pelkey, N. W. and Srinivas, V. (2009). 'Do mangroves provide an effective barrier to storm surges?'. *Proceedings of the National Academy of Sciences*, vol 106, no 40, p. E111. doi:https://doi.org/10.1073/pnas.09008799106.

Bao, T. Q. (2011). 'Effect of mangrove forest structures on wave attenuation in coastal Vietnam'. *Oceanologia*, vol 53, no 3, pp. 807–818.

Barbier, E. B. (2013). 'Valuing ecosystem services for coastal wetland protection and restoration: Progress and challenges'. *Resources*, vol 2, no 3, pp. 213–230. doi:10.3390/resources2030213.

Barbier, E. B., Hacker, S. D., Kennedy, C., Koch, E. W., Stier, A. C. and Silliman, B. R. (2011). 'The value of estuarine and coastal ecosystem services'. *Ecological Monographs*, vol 81, no 2, pp. 169–193. doi:10.1890/10-1510.1.

Borja, Á., Dauer, D. M., Elliott, M. and Simenstad, C. A. (2010). 'Medium- and long-term recovery of estuarine and coastal ecosystems: Patterns, rates and restoration effectiveness'. *Estuaries and Coasts*, vol 33, no 6, pp. 1249–1260. doi:10.1007/s12237-010-9347-5.

Borsje, B. W., van Wesenbeeck, B. K., Dekker, F., Paalvast, P., Bouma, T. J., van Katwijk, M. M. and de Vries, M. B. (2011). 'How ecological engineering can serve in coastal protection'. *Ecological Engineering*, vol 37, no 2, pp. 113–122. doi:10.1016/j.ecoleng.2010.11.027.

Bouma, T. J., van Belzen, J., Balke, T., Zhu, Z., Airoldi, L., Blight, A. J., Davies, A. J., Galvan, C., Hawkins, S. J., Hoggart, S. P. G., Lara, J. L., Losada, I. J., Maza, M., Ondiviela, B., Skov, M. W., Strain, E. M., Thompson, R. C., Yang, S., Zanuttigh, B., Zhang, L. and Herman, P. M. J. (2014). 'Identifying knowledge gaps hampering application of intertidal habitats in coastal protection: Opportunities & steps to take'. *Coastal Engineering*, vol 87, pp. 147–157. doi:10.1016/j.coastal eng.2013.11.014.

Buckley, M. L., Lowe, R. J., Hansen, J. E. and van Dongeren, A. R. (2016). 'Wave setup over a fringing reef with large bottom roughness'. *Journal of Physical Oceanography*, vol 46, no 8, pp. 2317–2333. doi:10.1175/JPO-D-15-0148.1.

Chang, C.-W., Liu, P. L. F., Mei, C. C. and Maza, M. (2017). 'Modeling transient long waves propagating through a heterogeneous coastal forest of arbitrary shape'. *Coastal Engineering*, vol 122, pp. 124–140.

Chatenoux, B. and Peduzzi, P. (2007). 'Impacts from the 2004 Indian Ocean tsunami: Analysing the potential protecting role of environmental features'. *Natural Hazards*, vol 40, no 2, pp. 289–304.

Christianen, M. J. A., van Belzen, J., Herman, P. M. J., van Katwijk, M. M., Lamers, L. P. M., van Leent, P. J. M. and Bouma, T. J. (2013). 'Low-canopy seagrass beds still provide important coastal protection services'. *PLoS ONE*, vol 8, no 5, p. e62413. doi:10.1371/journal.pone.0062413.

Claflin, S. B. and Webb, C. E. (2017). 'Surrounding land use significantly influences adult mosquito abundance and species richness in urban mangroves'. *Wetlands Ecology Management*, vol 25, no 331. doi:10.1007/s11273-016-9520-0.

Cochard, R., Ranamukhaarachchi, S. L., Shivakoti, G. P., Shipin, O. V., Edwards, P. J. and Seeland, K. T. (2008). 'The 2004 tsunami in Aceh and Southern Thailand: A review on coastal ecosystems, wave hazards and vulnerability'. *Perspectives in Plant Ecology, Evolution and Systematics*, vol 10, no 1, pp. 3–40.

Connell, S. D., Russell, B. D., Turner, D. J., Shepherd, S. A., Kildea, T., Miller, D., Airoldi, L. and Cheshire, A. (2008). 'Recovering a lost baseline: Missing kelp forests from a metropolitan coast'. *Marine Ecology Progress Series*, vol 360, pp. 63–72.

Costanza, R., d'Arge, R., de Groot, R., Farber, S., Grasso, M., Hannon, B., Limburg, K., Naeem, S., O'Neill, R. V., Paruelo, J., Raskin, R. G., Sutton, P. and van den Belt, M. (1997). 'The value of the world's ecosystem services and natural capital'. *Nature*, vol 387, pp. 253–260. doi:10.1038/387253a0.

Costanza, R., Perez-Maqueo, O., Martinez, M. L., Sutton, P., Anderson, S. J. and Mulder, K. (2008). 'The value of coastal wetlands for hurricane protection'. *Ambio*, vol 37, no 4, pp. 241–248.

Das, S. and Vincent, J. R. (2009). 'Mangroves protected villages and reduced death toll during Indian super cyclone'. *Proceedings of the National Academy of Sciences*, vol 106, no 18, pp. 7357–7360.

de Boer, W. F. (2007). 'Seagrass–sediment interactions, positive feedbacks and critical thresholds for occurrence: A review'. *Hydrobiologia*, vol 591, no 1, pp. 5–24. doi:10.1007/s10750-007-0780-9.

Duarte, C. M. (2002). 'The future of seagrass meadows'. *Environmental Conservation*, vol 29, no 2, pp. 192–206.

Dubi, A. and Tørum, A. (1995). 'Wave damping by kelp vegetation', in B. L. Edge (ed.), *Proceedings of the Twenty-Fourth Coastal Engineering Conference in Kobe, Japan*, pp. 142–156. doi:10.1061/9780784400890.012.

Elwany, M. H. S., O'Reilly, W. C., Guza, R. T. and Flick, R. E. (1995). 'Effects of Southern California kelp beds on waves'. *Journal of Waterway, Port, Coastal, and Ocean Engineering*, vol 121, no 2, pp. 143–150.

Fagherazzi, S. (2014). 'Coastal processes: Storm-proofing with marshes'. *Nature Geoscience*, vol 7, no 10, pp. 701–702. doi:10.1038/ngeo2262.

Falter, J. L., Lowe, R. J., Zhang, Z. and McCulloch, M. (2013). 'Physical and biological controls on the carbonate chemistry of coral reef waters: Effects of metabolism, wave forcing, sea level, and geomorphology'. *PLoS ONE*, vol 8, no 1, p. e53303. doi:10.1371/journal.pone.0053303.

Filbee-Dexter, K., Feehan, C. J. and Scheibling, R. E. (2016). 'Large-scale degradation of a kelp ecosystem in an ocean warming hotspot'. *Marine Ecology Progress Series*, vol 543, pp. 141–152.

Fonseca, M. S., Koehl, M. A. R. and Kopp, B. S. (2007). 'Biomechanical factors contributing to self-organization in seagrass landscapes'. *Journal of Experimental Marine Biology and Ecology*, vol 340, no 2, pp. 227–246.

Fritz, H. M., Borrero, J. C., Synolakis, C. E. and Yoo, J. (2006). '2004 Indian Ocean tsunami flow velocity measurements from survivor videos'. *Geophysical Research Letters*, vol 33, no 24, p. L24605. doi:10.1029/2006GL026784.

Gambi, M. C., Nowell, A. R. M. and Jumars, P. A. (1990). 'Flume observations on flow dynamics in Zostera marina (eelgrass) beds'. *Marine Ecology Progress Series*, vol 61, no 1–2, pp. 159–169.

Gedan, K. B., Silliman, B. R. and Bertness, M. D. (2009). 'Centuries of human-driven change in salt marsh ecosystems'. *Annual Review of Marine Science*, vol 1, no 1, pp. 117–141. doi:10.1146/annurev.marine.010908.163930.

Ghisalberti, M. (2009). 'Obstructed shear flows: Similarities across systems and scales'. *Journal of Fluid Mechanics*, vol 641, pp. 51–61. doi:10.1017/S0022112009992175.

Granata, T. C., Serra, T., Colomer, J., Casamitjana, X., Duarte, C. M. and Gacia, E. (2001). 'Flow and particle distributions in a nearshore seagrass meadow before and after a storm'. *Marine Ecology Progress Series*, vol 218, pp. 95–106.

Guannel, G., Arkema, K., Ruggiero, P. and Verutes, G. (2016). 'The power of three: Coral reefs, seagrasses and mangroves protect coastal regions and increase their resilience'. *PLoS ONE*, vol 11, no 7, p. e0158094. doi:10.1371/journal.pone.0158094.

Hansen, J. C. R. and Reidenbach, M. A. (2012). 'Wave and tidally driven flows in eelgrass beds and their effect on sediment suspension'. *Marine Ecology Progress Series*, vol 448, pp. 271–288.

Hansen, J. C. R. and Reidenbach, M. A. (2013). 'Seasonal growth and senescence of a Zostera marina seagrass meadow alters wave-dominated flow and sediment suspension within a coastal bay'. *Estuaries and Coasts*, vol 36, no 6, pp. 1099–1114.

Harada, K. and Imamura, F. (2005). 'Effects of coastal forest on tsunami hazard mitigation – A preliminary investigation', in K. Satake (ed.), *Tsunamis*. Dordrecht: Springer, pp. 279–292.

Harris, D. L., Rovere, A., Casella, E., Power, H., Canavesio, R., Collin, A., Pomeroy, A., Webster, J. M. and Parravicini, V. (2018). 'Coral reef structural complexity provides important coastal protection from waves under rising sea levels'. *Science Advances*, vol 4, no 2, p. eaao4350. doi:10.1126/sciadv.aao4350.

Hearn, C. J. (1999). 'Wave-breaking hydrodynamics within coral reef systems and the effect of changing relative sea level'. *Journal of Geophysical Research: Oceans*, vol 104, no C12, pp. 30007–30019. doi:10.1029/1999JC900262.

Henderson, S. M., Norris, B. K., Mullarney, J. C. and Bryan, K. R. (2017). 'Wave-frequency flows within a near-bed vegetation canopy'. *Continental Shelf Research*, vol 147, pp. 91–101.

Hertweck, G. and Liebezeit, G. (2002). 'Historic mussel beds (*Mytilus edulis*) in the sedimentary record of a back-barrier tidal flat near Spiekeroog Island, southern North Sea'. *Helgoland Marine Research*, vol 56, no 1, pp. 51–58.

Horstman, E. M., Dohmen-Janssen, C. M., Narra, P. M. F., van den Berg, N. J. F., Siemerink, M. and Hulscher, S. J. M. H. (2014). 'Wave attenuation in mangroves: A quantitative approach to field observations'. *Coastal Engineering*, vol 94, pp. 47–62.

Horstman, E. M., Lundquist, C. J., Bryan, K. R., Bulmer, R. H., Mullarney, J. C. and Stokes, D. J. (2018). 'The dynamics of expanding mangroves in New Zealand', in C. Makowski and C. W. Finkl (eds.), *Threats to Mangrove Forests*. Cham, Switzerland: Springer, pp. 23–51.

Howes, N. C., FitzGerald, D. M., Hughes, Z. J., Georgiou, I. Y., Kulp, M. A., Miner, M. D., Smith, J. M. and Barras, J. A. (2010). 'Hurricane-induced failure of low salinity wetlands'. *Proceedings of the National Academy of Sciences*, vol 107, no 32, pp. 14014–14019. doi:10.1073/pnas.0914582107.

Hughes, T. P., Kerry, J. T., Álvarez-Noriega, M., Álvarez-Romero, J. G., Anderson, K. D., Baird, A. H., Babcock, R. C., Beger, M., Bellwood, D. R., Berkelmans, R., Bridge, T. C., Butler, I. R., Byrne, M., Cantin, N. E., Comeau, S., Connolly, S. R., Cumming, G. S., Dalton, S. J., Diaz-Pulido, G., Eakin, C. M., Figueira, W. F., Gilmour, J. P., Harrison, H. B., Heron, S. F., Hoey, A. S., Hobbs, J.-P. A., Hoogenboom, M. O., Kennedy, E. V., Kuo, C.-Y., Lough, J. M., Lowe, R. J., Liu, G., McCulloch, M. T., Malcolm, H. A., McWilliam, M. J., Pandolfi, J. M., Pears, R. J., Pratchett, M. S., Schoepf, V., Simpson, T., Skirving, W. J., Sommer, B., Torda, G., Wachenfeld, D. R., Willis, B. L. and Wilson, S. K. (2017). 'Global warming and recurrent mass bleaching of corals'. *Nature*, vol 543, pp. 373–377. doi:10.1038/nature21707.

Jackson, G. A. (1984). 'Internal wave attenuation by coastal kelp stands'. *Journal of Physical Oceanography*, vol 14, no 8, pp. 1300–1306. doi:10.1175/1520-0485(1984)014<1300:IWABCK>2.0.CO;2.

Jackson, G. A. (1997). 'Currents in the high drag environment of a coastal kelp stand off California'. *Continental Shelf Research*, vol 17, no 15, pp. 1913–1928. doi:https://doi.org/10.1016/S0278-4343(97)00054-X.

Jackson, G. A. and Winant, C. D. (1983). 'Effect of a kelp forest on coastal currents'. *Continental Shelf Research*, vol 2, no 1, pp. 75–80.

John, B. M., Shirlal, K. G., Rao, S. and Rajasekaran, C. (2016). 'Effect of artificial seagrass on wave attenuation and wave run-up'. *International Journal of Ocean and Climate Systems*, vol 7, no 1, pp. 14–19.

Jones, H. P., Hole, D. G. and Zavaleta, E. S. (2012). 'Harnessing nature to help people adapt to climate change'. *Nature Climate Change*, vol 2, no 7, pp. 504–509. doi:10.1038/nclimate1463.

Kathiresan, K. and Bingham, B. L. (2001). 'Biology of mangroves and mangrove eco-systems'. *Advances in Marine Biology*, vol 40, pp. 81–251.

Kathiresan, K. and Rajendran, N. (2005). 'Coastal mangrove forests mitigated tsunami'. *Estuarine, Coastal and shelf science*, vol 65, no 3, pp. 601–606.

Katwijk, M. M., Thorhaug, A., Marbà, N., Orth, R. J., Duarte, C. M., Kendrick, G. A., Althuizen, I. H. J., Balestri, E., Bernard, G. and Cambridge, M. L. (2016). 'Global analysis of seagrass restoration: The importance of large-scale planting'. *Journal of Applied Ecology*, vol 53, no 2, pp. 567–578.

Kench, P. S. and Brander, R. W. (2006). 'Wave processes on coral reef flats: Implications for reef geomorphology using Australian case studies'. *Journal of Coastal Research*, vol 22, no 1, pp. 209–223. doi:10.2112/05A-0016.1.

Kilminster, K., McMahon, K., Waycott, M., Kendrick, G. A., Scanes, P., McKenzie, L., O'Brien, K. R., Lyons, M., Ferguson, A. and Maxwell, P. (2015). 'Unravelling complexity in seagrass systems for management: Australia as a microcosm'. *Science of the Total Environment*, vol 534, pp. 97–109.

Kobayashi, N., Raichle, A. W. and Asano, T. (1993). 'Wave attenuation by vegetation'. *Journal of Waterway, Port, Coastal, and Ocean Engineering*, vol 119, no 1, pp. 30–48.

Koch, E. W., Barbier, E. B., Silliman, B. R., Reed, D. J., Perillo, G. M. E., Hacker, S. D., Granek, E. F., Primavera, J. H., Muthiga, N. and Polasky, S. (2009). 'Nonlinearity in ecosystem services: Temporal and spatial variability in coastal protection'. *Frontiers in Ecology and the Environment*, vol 7, no 1, pp. 29–37.

Krauss, K. W., Doyle, T. W., Doyle, T. J., Swarzenski, C. M., From, A. S., Day, R. H. and Conner, W. H. (2009). 'Water level observations in mangrove swamps during two hurricanes in Florida'. *Wetlands*, vol 29, no 1, pp. 142–149.

Krumhansl, K. A., Okamoto, D. K., Rassweiler, A., Novak, M., Bolton, J. J., Cavanaugh, K. C., Connell, S. D., Johnson, C. R., Konar, B. and Ling, S. D. (2016). 'Global patterns of kelp forest change over the past half-century'. *Proceedings of the National Academy of Sciences*, vol 113, no 48, pp. 13785–13790.

Lawson, S. E., McGlathery, K. J. and Wiberg, P. L. (2012). 'Enhancement of sediment suspension and nutrient flux by benthic macrophytes at low biomass'. *Marine Ecology Progress Series*, vol 448, pp. 259–270.

Løvås, S. M. and Tørum, A. (2001). 'Effect of submerged vegetation upon wave damping and run-up on beaches: A case study on Laminaria hyperborea', in B. L. Edge (ed.), *Proceedings of the Twenty-Seventh Coastal Engineering Conference in Sydney, Australia*, American Society of Civil Engineers, pp. 851–864.

Lovelock, C. E., Feller, I. C., Reef, R., Hickey, S. and Ball, M. C. (2017). 'Mangrove dieback during fluctuating sea levels'. *Scientific Reports*, vol 7, no 1, p. 1680.

Lowe, R. J., Falter, J. L., Bandet, M. D., Pawlak, G., Atkinson, M. J., Monismith, S. G. and Koseff, J. R. (2005a). 'Spectral wave dissipation over a barrier reef'. *Journal of Geophysical Research: Oceans*, vol 110, no C4, pp. C04001. doi:10.1029/2004JC002711.

Lowe, R. J., Koseff, J. R. and Monismith, S. G. (2005b). 'Oscillatory flow through submerged canopies: 1. Velocity structure'. *Journal of Geophysical Research: Oceans*, vol 110, no C10. doi: 10.1029/2004JC002788.

Lowe, R. J., Falter, J. L., Koseff, J. R., Monismith, S. G. and Atkinson, M. J. (2007). 'Spectral wave flow attenuation within submerged canopies: Implications for wave energy dissipation'. *Journal of Geophysical Research: Oceans*, vol 112, no 5. doi:10.1029/2006JC003605.

Ludka, B. C., Guza, R. T., O'Reilly, W. C. and Yates, M. L. (2015). 'Field evidence of beach profile evolution toward equilibrium'. *Journal of Geophysical Research: Oceans*, vol 120, no 11, pp. 7574–7597. doi:10.1002/2015JC010893.

McDonald, C. B., Koseff, J. R. and Monismith, S. G. (2006). 'Effects of the depth to coral height ratio on drag coefficients for unidirectional flow over coral'. *Limnology and Oceanography*, vol 51, no 3, pp. 1294–1301.

McGranahan, G., Balk, D. and Anderson, B. (2007). 'The rising tide: Assessing the risks of climate change and human settlements in low elevation coastal zones'. *Environment and Urbanization*, vol 19, no 1, pp. 17–37.

McIvor, A. L., Möller, I. and Spencer, T. (2012). 'Reduction of wind and swell waves by mangroves'. Natural Coastal Protection Series: Report 1. Cambridge Coastal Research Unit Working Paper 40. The Nature Conservancy and Wetlands International, Cambridge, UK. Available at: www.conservationgateway.org/Conservation Practices/Marine/crr/library/Documents/wind-and-swell-wave-reduction-by-mangroves.pdf.

McKee, K. L., Cahoon D. R. and Feller I. C. (2007). 'Caribbean mangroves adjust to rising sea level through biotic controls on change in soil elevation'. *Global Ecology and Biogeography*, vol 16, no 5, pp. 545–556. doi:10.1111/j.1466-8238.2007.00317.x.

Mcowen, C. J., Weatherdon, L. V., Bochove, J.-W. V., Sullivan, E., Blyth, S., Zockler, C., Stanwell-Smith, D., Kingston, N., Martin, C. S., Spalding, M. and Fletcher, S. (2017). 'A global map of saltmarshes'. *Biodiversity Data Journal*, vol 5, p. e11764. doi:10.3897/BDJ.5.e11764.

Massel, S. R., Furukawa, K. and Brinkman, R. M. (1999). 'Surface wave propagation in mangrove forests'. *Fluid Dynamics Research*, vol 24, no 4, pp. 219–249.

Mazda, Y., Magi, M., Ikeda, Y., Kurokawa, T. and Asano, T. (2006). 'Wave reduction in a mangrove forest dominated by *Sonneratia* sp'. *Wetlands Ecology and Management*, vol 14, no 4, pp. 365–378.

Möller, I., Kudella, M., Rupprecht, F., Spencer, T., Paul, M., van Wesenbeeck, B. K., Wolters, G., Jensen, K., Bouma, T. J., Miranda-Lange, M. and Schimmels, S. (2014). 'Wave attenuation over coastal salt marshes under storm surge conditions'. *Nature Geoscience*, vol 7, pp. 727–731. doi:10.1038/ngeo2251.

Montakhab, A., Yusuf, B., Ghazali, A. H. and Mohamed, T. A. (2012). 'Flow and sediment transport in vegetated waterways: A review'. *Reviews in Environmental Science and Bio/Technology*, vol 11, no 3, pp. 275–287.

Morris, R. L., Konlechner, T. M., Ghisalberti, M. and Swearer, S. E. (2018). 'From grey to green: Efficacy of eco-engineering solutions for nature-based coastal defence'. *Global Change Biology*, vol 24, no 5, pp. 1827–1842. doi:10.1111/gcb.14063.

Narayan, S., Beck, M. W., Reguero, B. G., Losada, I. J., van Wesenbeeck, B., Pontee, N., Sanchirico, J. N., Ingram, J. C., Lange, G.-M. and Burks-Copes, K. A. (2016). 'The effectiveness, costs and coastal protection benefits of natural and nature-based defences'. *PLoS ONE*, vol 11, no 5, p. e0154735. doi:10.1371/journal.pone.0154735.

Nepf, H. M. (2012). 'Hydrodynamics of vegetated channels'. *Journal of Hydraulic Research*, vol 50, no 3, pp. 262–279.

Norris, B. K., Mullarney, J. C., Bryan, K. R. and Henderson, S. M. (2017). 'The effect of pneumatophore density on turbulence: A field study in a *Sonneratia*-dominated mangrove forest, Vietnam'. *Continental Shelf Research*, vol 147, pp. 114–127.

O'Donnell, J. E. D. (2016). 'Living shorelines: A review of literature relevant to New England coasts'. *Journal of Coastal Research*, vol 33, no 2, pp. 435–451.

Orth, R. J., Carruthers, T. J. B., Dennison, W. C., Duarte, C. M., Fourqurean, J. W., Heck, K. L., Hughes, A. R., Kendrick, G. A., Kenworthy, W. J. and Olyarnik, S. (2006). 'A global crisis for seagrass ecosystems'. *Bioscience*, vol 56, no 12, pp. 987–996.

Paling, E. I., Kobryn, H. T. and Humphreys, G. (2008). 'Assessing the extent of mangrove change caused by Cyclone Vance in the eastern Exmouth Gulf, north-western Australia'. *Estuarine, Coastal and Shelf Science*, vol 77, no 4, pp. 603–613. doi:https://doi.org/10.1016/j.ecss.2007.10.019.

Parvathy, K. G., Umesh, P. A. and Bhaskaran, P. K. (2017). 'Inter-seasonal variability of wind-waves and their attenuation characteristics by mangroves in a reversing wind system'. *International Journal of Climatology*, vol 37, no 15, pp. 5089–5106.

Péquignet, A. C. N., Becker, J. M., Merrifield, M. A. and Aucan, J. (2009). 'Forcing of resonant modes on a fringing reef during tropical storm Man Yi'. *Geophysical Research Letters*, vol 36, no 3, p. L03607. doi:10.1029/2008GL036259.

Perry, C. T., Spencer, T. and Kench, P. S. (2008). 'Carbonate budgets and reef production states: A geomorphic perspective on the ecological phase-shift concept'. *Coral Reefs*, vol 27, no 4, pp. 853–866.

Piazza, B. P., Banks, P. D. and La Peyre, M. K. (2005). 'The potential for created oyster shell reefs as a sustainable shoreline protection strategy in Louisiana'. *Restoration Ecology*, vol 13, no 3, pp. 499–506.

Pomeroy, A. W. M., Lowe, R. J., Ghisalberti, M., Storlazzi, C., Symonds, G. and Roelvink, D. (2017). 'Sediment transport in the presence of large reef bottom roughness'. *Journal of Geophysical Research: Oceans*, vol 122, no 2, pp. 1347–1368.

Quartel, S., Kroon, A., Augustinus, P., van Santen, P. and Tri, N. H. (2007). 'Wave attenuation in coastal mangroves in the Red River Delta, Vietnam'. *Journal of Asian Earth Sciences*, vol 29, no 4, pp. 576–584.

Renema, W., Pandolfi, J. M., Kiessling, W., Bosellini, F. R., Klaus, J. S., Korpanty, C., Rosen, B. R., Santodomingo, N., Wallace, C. C., Webster, J. M. and Johnson, K. G. (2016). 'Are coral reefs victims of their own past success?'. *Science Advances*, vol 2, no 4, p. e1500850. doi: 10.1126/sciadv.1500850.

Roeber, V. and Bricker, J. D. (2015). 'Destructive tsunami-like wave generated by surf beat over a coral reef during Typhoon Haiyan'. *Nature Communications*, vol 6, p. 7854. doi:10.1038/ncomms8854.

Rosman, J. H. and Hench, J. L. (2011). 'A framework for understanding drag parameterizations for coral reefs'. *Journal of Geophysical Research: Oceans*, vol 116, no C8, p. C08025. doi:10.1029/2010JC006892.

Rosman, J. H., Koseff, J. R., Monismith, S. G. and Grover, J. (2007). 'A field investigation into the effects of a kelp forest (*Macrocystis pyrifera*) on coastal hydrodynamics and transport'. *Journal of Geophysical Research: Oceans*, vol 112, no C2. doi:10. 1029/2005JC003430.

Rupprecht, F., Möller, I., Paul, M., Kudella, M., Spencer, T., van Wesenbeeck, B. K., Wolters, G., Jensen, K., Bouma, T. J., Miranda-Lange, M. and Schimmels, S. (2017). 'Vegetation-wave interactions in salt marshes under storm surge conditions'. *Ecological Engineering*, vol 100, pp. 301–315. doi:10.1016/j.ecoleng.2016.12.030.

Sanderson, P. G. and Eliot, I. (1996). 'Shoreline salients, cuspate forelands and tombolos on the coast of Western Australia' *Journal of Coastal Research*, vol 12, no 3, pp. 761–773.

Shepard, C. C., Crain, C. M. and Beck, M. W. (2011). 'The protective role of coastal marshes: A systematic review and meta-analysis'. *PLoS ONE*, vol 6, no 11, p. e27374. doi:10.1371/journal.pone.0027374.

Silliman, B. R., van de Koppel, J., McCoy, M. W., Diller, J., Kasozi, G. N., Earl, K., Adams, P. N. and Zimmerman, A. R. (2012). 'Degradation and resilience in Louisiana salt marshes after the BP-*Deepwater Horizon* oil spill'. *Proceedings of the National Academy of Sciences*, vol 109, no 28, pp. 11234–11239.

Small, C. and Nicholls, R. J. (2003). 'A global analysis of human settlement in coastal zones'. *Journal of Coastal Research*, vol 19, no 3, pp. 584–599.

Spalding, M., Ravilious, C. and Green, E. P. (2001). *World Atlas of Coral Reefs*. Berkeley, CA: University of California Press.

Spalding, M., McIvor, A. L., Beck, M. W., Koch, E. W., Möller, I., Reed, D. J., Rubinoff, P., Spencer, T., Tolhurst, T. J. and Wamsley, T. V. (2014). 'Coastal ecosystems: A critical element of risk reduction'. *Conservation Letters*, vol 7, no 3, pp. 293–301.

Spencer, T., Brooks, S. M. and Möller, I. (2014). 'Floods: Storm-surge impact depends on setting'. *Nature*, vol 505, no 7481, p. 26.

Spencer, T., Möller, I., Rupprecht, F., Bouma, T. J., Wesenbeeck, B. K., Kudella, M., Paul, M., Jensen, K., Wolters, G., Miranda-Lange, M. and Schimmels, S. (2015). 'Salt marsh surface survives true-to-scale simulated storm surges'. *Earth Surface Processes and Landforms*, vol 41, no 4, pp. 543–552. doi:10.1002/esp.3867.

Strusinska-Correia, A., Husrin, S. and Oumeraci, H. (2013). 'Tsunami damping by mangrove forest: A laboratory study using parameterized trees'. *Natural Hazards and Earth System Sciences*, vol 13, no 2, pp. 483–503.

Temmerman, S., Meire, P., Bouma, T. J., Herman, P. M. J., Ysebaert, T. and De Vriend, H. J. (2013). 'Ecosystem-based coastal defence in the face of global change'. *Nature*, vol 504, no 7478, pp. 79–83. doi:10.1038/nature12859.

Thomas, N., Lucas, R., Bunting, P., Hardy, A., Rosenqvist, A. and Simard, M. (2017). 'Distribution and drivers of global mangrove forest change, 1996–2010'. *PLoS ONE*, vol 12, no 6, p. e0179302. doi:10.1371/journal.pone.0179302.

Tsai, C.-P., Ying-Chi, C., Sihombing, T. O. and Chang, L. (2017). 'Simulations of moving effect of coastal vegetation on tsunami damping'. *Natural Hazards and Earth System Sciences*, vol 17, no 5, pp. 693–702.

Tucker, M. J. (1950). 'Surf beats: Sea waves of 1 to 5 min. period'. *Proceedings of the Royal Society of London A: Mathematical, Physical and Engineering Sciences*, vol 202, no 1071, pp. 565–573. doi:10.1098/rspa.1950.0120.

van de Koppel, J., van der Wal, D., Bakker, J. P. and Herman, P. M. J. (2005). 'Self-organization and vegetation collapse in salt marsh ecosystems'. *The American Naturalist*, vol 165, no 1, pp. E1–E12. doi:10.1086/426602.

van Katwijk, M. M., Bos, A. R., Hermus, D. C. R. and Suykerbuyk, W. (2010). 'Sediment modification by seagrass beds: Muddification and sandification induced by plant cover and environmental conditions'. *Estuarine, Coastal and Shelf Science*, vol 89, no 2, pp. 175–181.

van Rooijen, A. A., McCall, R. T., van Thiel de Vries, J. S. M., van Dongeren, A. R., Reniers, A. and Roelvink, J. A. (2016). 'Modeling the effect of wave–vegetation interaction on wave setup'. *Journal of Geophysical Research: Oceans*, vol 121, no 6, pp. 4341–4359.

van Slobbe, E., de Vriend, H. J., Aarninkhof, S., Lulofs, K., de Vries, M. and Dircke, P. (2013). 'Building with nature: In search of resilient storm surge protection strategies'. *Natural Hazards*, vol 66, no 3, pp. 1461–1480. doi:10.1007/s11069-013-0612-3.

Vo-Luong, P. and Massel, S. (2008). 'Energy dissipation in non-uniform mangrove forests of arbitrary depth'. *Journal of Marine Systems*, vol 74, no 1–2, pp. 603–622.

Vu, M. T., Lacroix, Y. and Nguyen, V. T. (2017). 'Investigating the impacts of the regression of Posidonia oceanica on hydrodynamics and sediment transport in Giens Gulf'. *Ocean Engineering*, vol 146, pp. 70–86. doi:https://doi.org/10.1016/j.oceaneng.2017.09.051.

Wahl, T., Haigh, I. D., Nicholls, R. J., Arns, A., Dangendorf, S., Hinkel, J. and Slangen, A. B. A. (2017). 'Understanding extreme sea levels for broad-scale coastal impact and adaptation analysis'. *Nature Communications*, vol 8, p. 16075. doi:10.1038/ncomms16075.

Wolters, M., Bakker J. P., Bertness M. D., Jefferies R. L. and Möller, I. (2005). 'Salt-marsh erosion and restoration in south-east England: Squeezing the evidence requires realignment'. *Journal of Applied Ecology*, vol 42, no 5, pp. 844–851. doi:10.1111/j.1365-2664.2005.01080.x.

Wright, L. D., May, S. K., Short, A. D. and Green, M. O. (1984). 'Beach and surf zone equilibria and response times', in B. L. Edge (ed.), *Proceedings of the Nineteenth International Conference on Coastal Engineering in Houston, Texas*, American Society of Civil Engineers, pp. 2150–2164.

Young, I. R., Zieger, S. and Babanin, A. V. (2011). 'Global trends in wind speed and wave height'. *Science*, vol 332, no 6028, pp. 451–455.

Zhang, K., Liu, H., Li, Y., Xu, H., Shen, J., Rhome, J. and Smith III, T. J. (2012). 'The role of mangroves in attenuating storm surges'. *Estuarine, Coastal and Shelf Science*, vol 102, pp. 11–23.

Part III

Healthy oceans

6 Monitoring ocean and estuary health

Paul G Thomson, Belinda L Cannell,
Anas Ghadouani, Matthew W Fraser and
William J Rayment

Introduction

The global oceans contribute significantly to humankind through the blue economy which can be defined as the sustainable use of the oceans for economic growth, improved livelihoods and jobs and ocean ecosystem health (World Bank and United Nations Department of Economic and Social Affairs, 2017). The oceans are central to our way of life, whether it be directly through socio-economic needs such as providing food, employment, transport and recreation, or indirectly by modulating our climate. While we rely heavily on our oceans, we also impact them through habitat destruction and pollution of our estuaries via run-off from urban, agricultural and industrial sources and activities such as resource extraction. Furthermore, it is now clear that we are impacting the oceans heavily through climate change, which is leading to increased seawater temperatures, changes in ocean circulation patterns and ocean acidification. The consequences of these changes include extreme marine events such as marine heatwaves, coral bleaching, rising sea levels and the increasing frequency and intensity of storms (Hoegh-Guldberg and Bruno, 2010). A pressing twenty-first century question, therefore, is: 'How do we respond to these changes?'.

The first step in response to this question is to understand how our estuaries, coastal seas and oceans are changing. Over the past 25 years, in particular, there has been a global push to monitor ocean health through long term observations of physical, geochemical and biological properties. Global, regional and national observing systems are now building baseline datasets using essential ocean variables such as seawater temperature and salinity and the distribution and abundance of plankton, fish, birds and mammals. These essential ocean variables, born from physical and biological indicators in use for decades, will be crucial in understanding gradual change in our oceans and the impacts of extreme marine events. Here we explore ocean monitoring efforts on global and national scales with examples from the island nations of Australia and New Zealand. Finally, we discuss the use of essential ocean variables and biological indicators for monitoring long and short term or extreme changes in our oceans (an example of which is given in Chapter 8).

Monitoring systems

Global monitoring

Global monitoring is led by the Global Ocean Observing System (GOOS), a programme of the Intergovernmental Oceanographic Commission (IOC) that is co-sponsored by the World Meteorological Organisation (WMO), the International Council for Science (ICSU) and the United Nations Environment Programme (UNEP). GOOS was established in the early 1990s to coordinate continuous, long term observations on the physical, chemical and biological properties of ocean and estuarine environments and make this data freely available to all interested parties (governments, industries, scientists, etc.) (Malone and Cole, 2000). Ocean health is one of three main themes of GOOS, defined as 'the ability of marine ecosystems to thrive and support human livelihoods' and is 'broadly based on factors that affect productivity, species diversity, and resilience such as adaptation to climate change' (GOOS, 2018). The GOOS programme contributes to ocean health studies by monitoring biodiversity and habitat, ocean acidification and water quality through the collection of essential ocean variables (EOVs) such as sea surface and subsurface water temperature, salinity, sea level, sea ice, currents, nutrient and carbon dioxide concentrations, ocean colour and phytoplankton distribution (Figure 6.1) (Constable *et al.*, 2016; Miloslavich *et al.*, 2018). The ocean health theme is supported by data from the themes of climate, measuring ocean heat content and sea levels and operational ocean services, which supports early warning systems for ocean hazards and ocean forecasting. The objectives of GOOS are implemented by the IOC (www.unesco.org/new/en/natural-sciences/ioc-oceans/) through collaboration with the GOOS regional alliances (GRAs) of national organisations which cover most of the world's oceans. The existing 13 GRAs have become major contributors of research and data to the global system and, in the southern hemisphere, include the Australian Integrated Marine Observing System (IMOS) and the Pacific Island GOOS (PI-GOOS).

Monitoring systems in Australia and New Zealand

The island nations of Australia and New Zealand are in intimate contact with their surrounding oceans; 85 per cent and 65 per cent respectively of the populations of both countries live near coastlines (Clark and Johnston 2017; Stats NZ 2016) and both rely heavily on the ocean for trade, fishing, transport and recreation. In Australia, the importance of monitoring variability and change in the surrounding oceans was recognised through the establishment of IMOS in 2007 (http://imos.org.au/). IMOS is a partnership between Australia's marine research institutes, government bodies and universities and is funded by the Federal Government (Hill *et al.*, 2010). IMOS was developed to establish a multidisciplinary observing system around Australia, monitoring

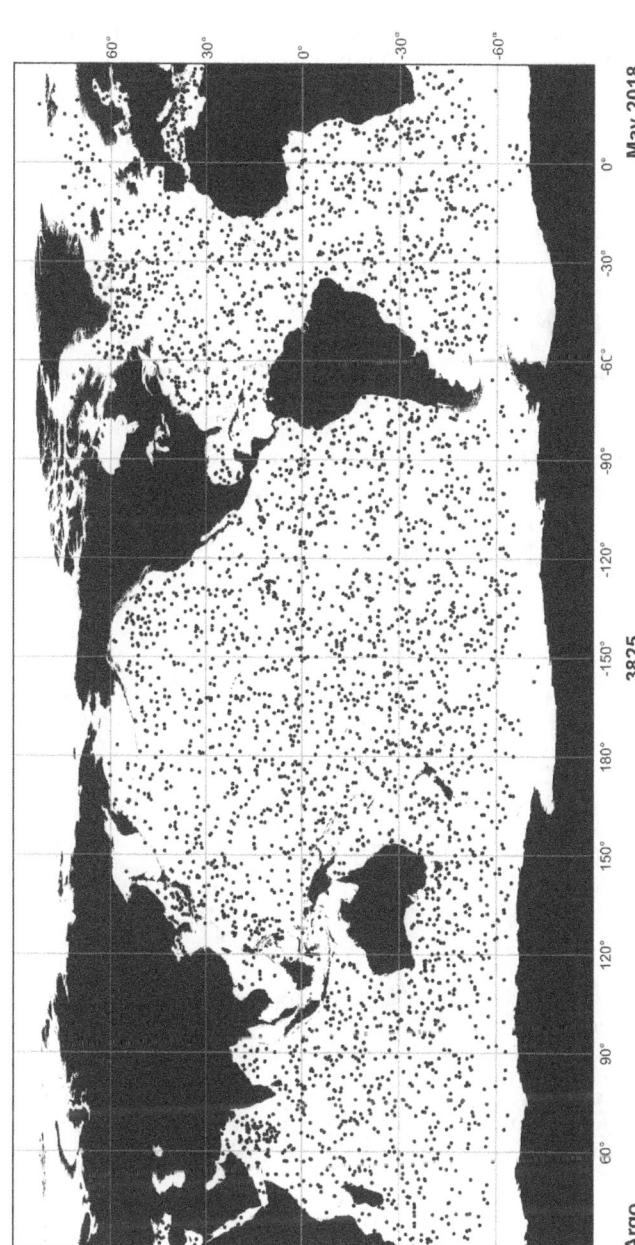

Argo
Latest location of operational floats (data distributed within the last 30 days)

3825

May 2018

Figure 6.1 The Global Ocean Observing System (GOOS).

Source: generated by www.jcommops.org on 6 June 2018.

Note
GOOS coordinates a collaborative system of ocean observations such as the Argo float programme, a global array of > 3,800 free-drifting profiling floats that measures the temperature, salinity and biogeochemistry of the upper 2,000m of the ocean. Each dot plotted above is the position of an Argo float on 6 June 2018.

the oceans using infrastructure such as satellite remote sensing, moorings, gliders, floats, ships of opportunity, biogeochemical sampling and an animal tracking network (Figure 6.2) (Hill *et al.*, 2010; Hoenner *et al.*, 2018; Lynch *et al.*, 2014; Pattiaratchi *et al.*, 2017). IMOS data is freely available from the Australian Ocean Data Network portal (AODN) (https://portal.aodn.org.au/)

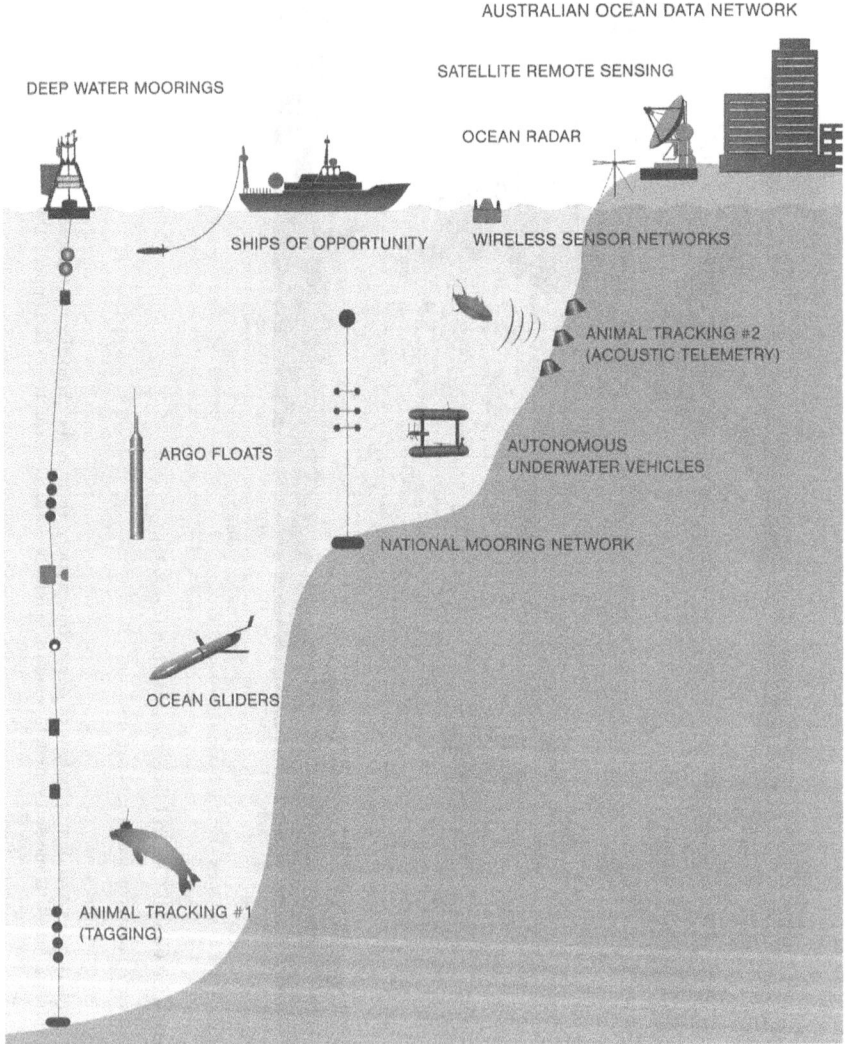

Figure 6.2 Integrated Marine Observing System (IMOS).

Source: IMOS.

Note
All IMOS data is freely available from the Australian Ocean Data Network (https://portal.aodn.org.au/).

and to encourage data uptake, several papers have been published detailing data availability, quality control techniques and examples of how the portal can be used (Davies *et al.*, 2016; Davies *et al.*, 2018; Hanson *et al.*, 2016; Hoenner *et al.*, 2018).

Long term monitoring for oceanographic and ecological change in Australia is enabled through the IMOS National Reference Stations (NRS), a series of moorings extending across 30 degrees of latitude in different biological ocean provinces (Lynch *et al.*, 2014). Like other moorings, these sites measure physical seawater parameters such as temperature, conductivity (salinity), ocean currents, and dissolved oxygen and turbidity concentrations. These measurements are combined with monthly biogeochemical samples at the mooring sites for carbon, nutrients, chlorophyll *a*, and bacteria, phytoplankton and zooplankton concentrations. Additionally, satellite remote sensing has been extremely useful when paired with the *in situ* NRS data, placing the observations in a larger regional and three-dimensional context. For example, monitoring sea surface temperatures, currents and chlorophyll concentrations extended the geographic context of marine heatwaves at two NRS locations in recent studies (Oliver *et al.*, 2018; Thompson *et al.*, 2015).

While most NRS are relatively new (since 2007), the network incorporates long term sites such Maria Island in Tasmania, Port Hacking near Sydney in New South Wales and Rottnest Island near Perth in Western Australia. The ability of these sites to monitor change in the coastal oceans is best illustrated by data from the Maria Island NRS on the east coast of Tasmania. The Maria Island NRS, managed by the Commonwealth Scientific and Industrial Research Organisation (CSIRO), has been recording data on seawater temperature and salinity since 1944. To date, this site has detected seawater warming occurring at a rate of four times the global average due to the increasing strength of the East Australian Current extension and its eddies that impact the east coast of Tasmania (Oliver *et al.*, 2017; Ridgway, 2007). In fact, this region is now recognised as a global warming hotspot that is experiencing unprecedented marine heatwaves (Oliver *et al.*, 2017; Pecl *et al.*, 2014). As a direct result of this monitoring, ecological impacts can be linked to increasing seawater temperatures in the region. These impacts include a decrease in phytoplankton growth rates and biomass and changes in zooplankton community composition, southward range extensions of macroinvertebrates and fish, loss of kelp forests and impacts on wild abalone and cultured species such as Atlantic salmon (Johnson *et al.*, 2011; Oliver *et al.*, 2017; Thompson *et al.*, 2009).

The longest running set of oceanographic measurements in New Zealand is the Portobello sea surface temperature (SST) time-series. Daily SST data have been gathered consistently at the Portobello Marine Laboratory in Otago Harbour, southeast New Zealand since 1953 (Greig *et al.*, 1988; Shaw *et al.*, 1999). Recent analyses of these data have revealed a warming trend of +0.10°C per decade between 1953 and 2016 (Shears and Bowen, 2017). A suite of oceanographic parameters has been consistently measured on the

Munida transect (Currie *et al.*, 2011), a shipboard time-series based on the transect studied by Jillett (1969). The 60 km transect runs offshore from Otago Peninsula, across the subtropical convergence, and into sub-Antarctic surface water. Among other things, the time-series data have been used to study temporal variation in carbon dioxide drawdown (Currie *et al.*, 2011) and iron speciation (Sander *et al.*, 2015) in surface waters.

Monitoring in estuaries

While much information is being gathered for a number of parameters within the ocean, these parameters are not only dependent on events occurring within the ocean space. In fact, ocean dynamics are also defined through their critical interaction with coastal and estuarine systems (as explored by Winter *et al.* in Chapter 5). There are more than 1,000 estuaries along the entire Australian coastline; i.e. approximately one estuarine system (inlet) every 34 km. Despite the environmental, cultural and economic value of these systems, they remain largely under-monitored; in addition, significant national monitoring programmes (e.g. IMOS) have not yet been expanded to include estuaries or, at least, some estuaries of significance. The lack of a national monitoring programme at this stage is due to a lack of resources, but is also a result of artificial politico-legal (state) boundaries that do not match ecosystems or oceanographic processes. The notable decline in water quality in many estuaries in Australia (Hamilton, 2000; Hamilton *et al.*, 2001), and around the world, is a clear indication of the significant impact of human activities, which is projected to increase in Australia as the population continues to grow towards doubling by 2050. (See Chapter 7 for further discussion of these impacts, especially from land-based sources of pollution.) While some estuaries of significance, such as the Swan–Canning system in Western Australia, have been monitored regularly over the past 20 years or so for basic water quality variables, those monitoring programmes have lacked the spatial coverage necessary for process understanding. For example, phytoplankton dynamics have been shown to be closely controlled by a suite of drivers including rainfall and hydrological processes in the catchment, which have been monitored, as well as estuarine physical dynamics such as salt-wedge movement and tidal diurnal cycles (Twomey and John, 2001). However, these latter physical processes have not been included in the monitoring programme. Overall, the need for high resolution monitoring will become more critical in the future, as will be the need for fully integrated monitoring programmes to include more coastal systems and estuaries.

Biological indicators for monitoring long and short term change

GOOS has recommended a range of essential ocean variables (EOVs) to monitor long term change (Miloslavich *et al.*, 2018). The EOVs are divided into two categories. The first category represents ecosystem components, from the biomass and diversity of phytoplankton and zooplankton through to the distribution and abundance of fish, birds and mammals. The second category represents the health of the ecosystem and extent of the habitat such as the cover and composition of seagrass and macroalgae (Miloslavich *et al.*, 2018). While the concept of EOVs has been evolving over time, many species or groups of species from the EOV categories have long been used as indicators of short term changes resulting from extreme marine events, as well as longer term changes stemming from these events. Essentially, an effective indicator is one in which we: (1) understand the factors that cause variation within the indicator, i.e. the effect of background 'noise' within the system; and (2) have a known and sensitive response to changes in the variable that is being assessed (Furness *et al.*, 1993). As such, certain indicators will only be appropriate for detection of (1) specific types of change; (2) within specific spatial extents; and/or (3) within specific time frames. Detailed knowledge of the general ecology of the indicator is paramount, as well as the ease of obtaining the response data. Some indicators can be used for both short term and long term change, while others are best suited to one or the other. Below we give examples of how indicators have been or can be used to monitor both short term (extreme) and long term change in the marine environment.

Phytoplankton biomass and diversity

Phytoplankton biomass and diversity is a GOOS EOV recommended for monitoring change. Phytoplankton are good indicators of short term change or extreme events in the marine environment for several reasons. Their small size can afford them a large surface area to volume ratio (Raven, 1998). As a result, their internal organelles that regulate the cells' function (such as the nucleus, cytoplasm, mitochondria and chloroplasts) are in very close proximity to the surrounding seawater which can be advantageous; for example, they can scavenge for nutrients very efficiently and rapidly grow and divide under favourable conditions. Conversely, their small size also means they can be impacted very quickly by adverse physicochemical changes in their environment. Additionally, phytoplankton can also be dispersed long distances in currents or eddies in water favourable to their growth and survival (McManus and Woodson, 2012). As water entrained in currents often differs from the surrounding water mass they have penetrated, phytoplankton may be good biological tracers or indicators of this change (Thompson *et al.*, 2015).

The picoplankton, the smallest size fraction of the phytoplankton with a diameter of 0.2–2 µm, are good indicators of environmental stress or structural changes in an ecosystem (Lu *et al.*, 2009; Munawar and Weisse, 1989). The picoplankton include the cyanobacteria *Prochlorococcus* sp. (0.6 µm dia.) and *Synechococcus* sp. (1 µm dia.) and the picoeukaryotes, including the species *Micromonas*, for example. These cells can be ideal indicator species due to their small size, rapid division rates, their distribution and their specific environmental requirements. *Prochlorococcus* sp. is found in clear, nutrient poor, tropical waters (Partensky *et al.*, 1999a). In contrast, *Synechococcus* sp. is most abundant in nutrient rich areas of oceanic upwelling and coastal zones from the tropics to the subpolar regions (Partensky *et al.*, 1999b). When monitoring for these species, changes in the abundance of the picoplankton can be indicative of long term change or extreme events. For example, pico-plankton abundance is very sensitive to increased nutrient and contamination inputs in coastal zones and salt lakes, making them good markers of long term impacts or one-off extreme pollution events (Lu *et al.*, 2009; Munawar and Weisse, 1989). Furthermore, the picoplankton are predicted to increase in abundance globally with climate change impacts (Flombaum *et al.*, 2013) and, along Australian coastlines, IMOS data shows there appears to be a poleward shift in their distribution southwards due to strengthening boundary currents (Thompson *et al.*, 2009; Thompson *et al.*, 2015). These data also show that the picoplankton are markers of seasonal boundary current cycles and for extreme events such as marine heatwaves (see Chapter 8 for further details) (Thompson *et al.*, 2015; Thomson and Pattiaratchi, 2018).

Molecular indicators

Molecular indicators can also provide information on species' responses to environmental stress over short to medium timescales (minutes–weeks). Extreme events such as marine heatwaves pose a particular threat to sessile benthic organisms such as corals, seagrasses and seaweeds given their inability to move to a refuge area within thermal tolerance levels. Monitoring of these organisms – which often form the foundation of coastal ecosystems – typically relies on measuring morphological indicators such as percentage cover or shoot density to detect stress. However, by the time declining trends are detected using these methodologies, mortality has already occurred and options for intervention are limited (Davey *et al.*, 2016; Macreadie *et al.*, 2014). Molecular-based techniques are increasingly seen as a method that can quickly and effectively detect stress in benthic, habitat-forming organisms, in which specific genes that are over- or under-expressed after exposure to a stressor can be identified. For example, RNA sequencing (RNA-seq) has been used to show differential gene expression in seagrass, coral and seaweed tissues, providing molecular markers for conservation efforts (Davey *et al.*, 2017; Kong *et al.*, 2014; Libro *et al.*, 2013; Liu *et al.*, 2014; Salo *et al.*, 2015), which can be directly linked to management decisions. Similarly, there is growing recognition that the microbiome that is associated with a host organ-ism (collectively termed the 'holobiont') plays a central role in determining how marine organisms respond to extreme events, and molecular methods play a critical role in determining the composition of host microbiomes. For example, metagenomic analysis of coral microbiomes show that changes in environmental conditions such as elevated temperature and nutrients leads to a fundamental shift in the taxonomic and functional composition of the coral-associated microbiome (Thurber *et al.*, 2009). Understanding the inter-actions between marine hosts, microbial communities and the environment will be critical in understanding how habitat forming marine organisms respond to marine extremes (Mera and Bourne, 2018).

Molecular markers can also be useful for tracking changes in mobile marine organisms. Sampling of environmental DNA (eDNA) is increasingly being used to detect the distribution and biomass of ecologically and eco-nomically important marine organisms, avoiding biases associated with other methods such as underwater visual censuses, fishing surveys or historical catch data (Rees *et al.*, 2014; Yamamoto *et al.*, 2016). Using eDNA to track species distribution and biomass involves the collection of a small amount of seawater (<1 ml); species-specific eDNA can then be detected using quantitative poly-merase chain reaction (q-PCR). The quantity of eDNA in a sample is related to the presence/absence of a species and its abundance, allowing the distribu-tion and biomass of targeted species to be collected. This technique can be targeted at individual species of interest. For example, eDNA has been used to detect the distribution and biomass of economically important fisheries species such as Japanese Jack Mackerel (*Trachurus japonicus*) (Yamamoto *et al.*, 2016), and species of conservation interest such as the endangered tidewater goby (*Eucyclogobium newberryi*) (Schmelzle and Kinziger, 2016) or whale sharks (Sigsgaard *et al.*, 2017). Projects using eDNA can also be targeted to assess community diversity across several families of fish (e.g. Stat *et al.*, 2017; Thomsen *et al.*, 2016). eDNA has even been used for early detection of inva-sive species in new areas (Ardura and Planes, 2017; Borrell *et al.*, 2017). Given that species redistribution is predicted to become a major feature associated with extreme events such as marine heatwaves, molecular tools such as eDNA are likely to become important methods that can assist in the management of our marine ecosystems and resources in the future.

Seabirds

Seabirds are sensitive indicators of many aspects of environmental change as their ecology has often been well-studied, they tend to be high in the food chain, their colonial nesting ensures that large numbers can be sampled at a given time and they are long lived (Furness *et al.*, 1993; Furness and Cam-phuysen 1997). They have been used to monitor various forms of anthropo-genic extreme events, such as dredging, oil spills, chemical contamination (i.e. chemicals at a concentration above natural background or level of detection), pollution (the level of pollutants that cause harm) and marine plastic pollu-tion. However, for seabirds to be effective indicators of ecosystem change, we must be able to quantify expected variation in the seabird parameter, or suite of parameters. Furthermore, it is necessary to understand how the vari-ation changes both spatially and temporally (Piatt *et al.*, 2007).

Acute oil pollution has highly visible impacts on the mortality of seabirds. However, monitoring the breeding cycle of seabirds, in terms of timing, participation and success, can also indicate indirect impacts of oil pollution on prey stocks (Velando *et al.*, 2005). Furthermore, as each oil has its own chemical signature, it is possible to analyse and fingerprint the oil on the feathers of oiled seabirds and then identify the source of contamination, a process known as petroleomics (Henkel *et al.*, 2014; Sydnes 2017).

The increasing contamination of oceans and estuaries, and hence the marine food chain, with heavy metals, polyaromatic hydrocarbons, organochlorines and emerging contaminants such as phthalates and pharmaceutical products threatens the health of marine fauna and people. Monitoring contaminants in feathers, tissues and blood from seabirds can be used to identify the source and ongoing impacts of chemical contaminants, and the persistence of chemicals in the marine system. For example, the concentration of heavy metals in the feathers of European shag (*Phalacrocorax aristotelis*) chicks significantly increased following the 'Prestige' oil spill, which was also contaminated with heavy metals (Moreno *et al.*, 2011). Sampling chick feathers offers some advantages over sampling adult feathers when investigating extreme events. This is because they represent a specific time and spatial extent, with heavy metal levels in feathers obtained from the food resources fed by their parents (Furness 1993; Moreno *et al.*, 2011). As such, issues associated with bioaccumulation and broad-ranging areas of adults in interpreting the source, uptake and potential effect of contamination are reduced (Furness 1993).

Plastic pollution in the oceans and estuaries from both land-based and ocean-based sources has been identified for decades (Farrington and Takada, 2014). It can end up on beaches, remain circulating in the oceans for decades (Elliot and Elliot, 2013) or be ingested by marine fauna, including seabirds. In fact, the frequency of plastic ingestion by seabirds is used to indicate the condition of a marine system, such as the North Sea (OSPAR Commission, 2008). International efforts designed to reduce plastic pollution do not appear to be effective, as exemplified by the decadal increase in the proportion of fledgling flesh-footed shearwaters (*Puffinus carneipes*) with ingested plastic (Lavers *et al.*, 2014).

Seabirds can also be used to indicate changes in marine prey stocks in specific areas. For example, a reduction in the breeding success of kittiwakes (*Rissa tridactyla*) in Shetland was correlated with a reduction in the abundance of sandeels (Furness, 2007). Seabirds can indicate the efficacy of different fishery management practices, as demonstrated by the improved breeding success of kittiwakes following the closure of the sandeel fishery in the North Sea. The improvement was related to the birds feeding on older fish, which would have otherwise been taken by the fishery (Daunt *et al.*, 2008).

As indicated in the examples above, there are various seabird life history and physical parameters that can be used to assess the impact of environmental changes. However, different parameters can reflect change over varying temporal and spatial scales (Furness *et al.*, 2003). Changes in population size tend to reflect changes over years to decades and integrate activities over large areas, including migratory ranges. For example, adult southern black-browed albatrosses (*Thalassarche melanophrys*) breed in the austral spring and migrate to wintering grounds following breeding. Their survival increased with average sea surface temperatures when incubating eggs, but decreased when tuna longlining effort increased in their wintering zone (Rolland *et al.*, 2008). Timing of egg lay, breeding success and growth of chicks reflect changes within a breeding season, exemplified by the reduced growth rate of crested tern chicks raised during periods of low sardine abundance (McLeay *et al.*, 2009). Changes in foraging behaviour and diet composition reflect changes over hours to days, such as the increased distance travelled to foraging grounds by Peruvian boobies as a result of prey depletion caused by commercial fisheries (Bertrand *et al.*, 2012).

It is thus important to use informed and appropriate parameters based on the type of change that is being assessed. This is made even more challenging given that the response measured can also differ not only between species but between locations for the same species. By combining many parameters together, such as several breeding responses from multiple species at one site, or single species at multiple sites, it is possible to investigate ecosystem-wide based impacts. For example, the abundance of 23 focal species of seabirds per season in southern California coupled with changing abundance of forage fish and krill was correlated with changes in oceanographic variables (Sydeman *et al.*, 2015). More recently, the few existing multi-decadal seabird monitoring and research data sets have enabled monitoring of the effects of changing climatic variables (Cannell *et al.*, 2012; Chambers *et al.*, 2014). However, much longer time-series of data are critical to gaining better insights into the association between anthropogenic climate change and the effects on seabirds and the marine ecosystem.

Cetaceans

Similar to seabirds, our detailed knowledge of the ecology of many cetacean species makes them useful indicators of environmental change. Furthermore, their longevity and often extensive energy reserves means that fine-scale temporal changes are dampened, allowing medium term and long term trends to be measured. For these reasons, cetaceans have been described as 'ecosystem sentinels' (Moore, 2008). The effects of lethal impacts such as hunting and bycatch in fishing gear can be measured through time-series of demographic parameters. For example, several species of large whales are now recovering and recolonising former habitats following the cessation of commercial whaling (e.g. southern right whales, *Eubalaena australis*) (Carroll *et al.*, 2014; Jackson *et al.*, 2016); and a reduction in gillnet fishing in part of the range of Hector's dolphin (*Cephalorhynchus hectori*), has resulted in an increase in survival rate (Gormley *et al.*, 2012). More subtle changes in ecosystems can be revealed through changes in behaviour, distribution or reproductive output. The effects of climate change in the northern Pacific Ocean have resulted in delayed migration and expanded feeding ranges of grey whales (*Eschrichtius robustus*; reviewed in Moore, 2008), while variations in calving success of southern right whales have been linked to climate mediated variability in populations of their prey (Leaper *et al.*, 2006; Seyboth *et al.*, 2016). As technology advances, our ability to measure these changes improves. For example, the use of unmanned aerial vehicles ('drones') in cetacean research is allowing accurate measurements of body condition of large whales (e.g. Christiansen *et al.*, 2018), which can potentially be linked to variations in productivity of their foraging habitat. Furthermore, remote sensing of whales using satellite imagery in isolated areas such as the Kimberly in North West Australia may prove to be a cost effective long term monitoring method for humpback whales, *Megaptera novaeangliae* (Thums *et al.*, 2018).

Conclusions

The last 70 years have seen greater attention being given to monitoring our oceans. Long term monitoring sites such as the Portobello time-series in New Zealand and stations such as Maria Island in Tasmania, Australia now form the backbone of national efforts to monitor the health of the oceans. At a global level, the establishment of GOOS demonstrates a level of international cooperation on ocean observation unprecedented in most other contexts. Additionally, monitoring essential ocean variables and indicator species is essential as they can give us insights into the ecological impacts of physical changes in the oceans, especially in the context of climate change where impacts are likely to dramatically change the ocean environment. Furthermore, biological indicators can provide us with the opportunity to react to changes in a timely and appropriate manner. However, continued funding of these programmes and heeding the advice emanating from the studies are critical if we are to address the challenges facing the ocean and continue to reap the benefits we enjoy from it.

Our analysis has explored some of the monitoring systems that have emerged, yet there are many others implemented in different geographic locations and for a variety of purposes utilising new and emerging technologies (Macreadie *et al.*, 2018). Future programmes should also include more coastal and estuarine systems in order to achieve a broader and well-integrated understanding of ocean dynamics, especially in light of changes projected to affect the oceans at global, regional and local scales.

Monitoring is of considerable importance to a broad range of stakeholders. For example:

• Industries such as oil and gas, desalination, offshore energy and others along coastlines and estuaries require environmental baselines against which the relative impacts of industrial developments can be properly and fairly assessed.
• Society in general (and industry) require baselines to inform discussion and debate around industries' social licence to operate.

- Government agencies tasked with enhancing ocean governance require a detailed understanding of the current and predicted impacts of anthropogenic activities to design effective conservation and management regimes.
- Environmental law enforcement requires data acquired from monitoring to demonstrate regulatory breaches as a result of human activities.
- In the context of climate change, monitoring can provide the data needed for accurate predictions so that law- and policy-makers can design effective and efficient systems that prevent harm to the ocean, marine living resources and habitats.

Continued and improved monitoring is thus an inherent requirement for future-proofing the sustainable use of the ocean.

References

Ardura A and Planes S (2017). 'Rapid assessment of non-indigenous species in the era of the eDNA barcoding: A Mediterranean case study'. *Estuarine, Coastal and Shelf Science*, vol 188, pp. 81–87.

Bertrand S, Joo R, Arbulu Smet C, Tremblay Y, Barbraud C and Weimerskirch H (2012). 'Local depletion by a fishery can affect seabird foraging'. *Journal of Applied Ecology*, vol 49, no 5, pp. 1168–1177.

Borrell YJ, Miralles L, Do Huu H, Mohammed-Geba K, Garcia-Vazquez E (2017). 'DNA in a bottle – Rapid metabarcoding survey for early alerts of invasive species in ports'. *PLoS ONE*, vol 12, no 9, p. e0183347.

Cannell BL, Chambers LE, Wooller RD and Bradley JS (2012). 'Poorer breeding by little penguins near Perth, Western Australia is correlated with above average sea surface temperatures and a stronger Leeuwin Current'. *Marine & Freshwater Research*, vol 63, pp. 914–925.

Carroll EL, Rayment WJ, Alexander AM, Baker CS, Patenaude NJ, Steel D, Constantine R, Cole R and Boren LJ (2014). 'Reestablishment of former wintering grounds by New Zealand southern right whales'. *Marine Mammal Science*, vol 30, pp. 206–220.

Chambers LE, Dann P, Cannell B and Woehler EJ (2014). 'Climate as a driver of phenological change in southern seabirds. *International Journal of Biometeorology*, vol 58, no 4, pp. 603–612.

Christiansen F, Vivier F, Charlton C, Ward R, Auerson A, Burnell S and Bejder L (2018). 'Maternal body size and condition determine calf growth rates in southern right whales'. *Marine Ecology Progress Series*, vol 92, pp. 267–282.

Clark GF and Johnston EL (2017). 'Australia state of the environment 2016: Coasts'. Independent report to the Australian Government Minister for Environment and Energy, Australian Government Department of the Environment and Energy, Canberra.

Constable AJ, Costa DP, Schofield O, Newman L, Urban ER, Fulton EA, Melbourne-Thomas J, Ballerini T, Boyd PW, Brandt A, de la Mare WK, Edwards M, Eléaume M, Emmerson L, Fennel K, Fielding S, Griffiths H, Gutt J, Hindell MA, Hofmann EE, Jennings S, La HS, McCurdy A, Mitchell BG, Moltmann T, Muelbert M, Murphy E, Press AJ, Raymond B, Reid K, Reiss C, Rice J, Salter I,

Smith DC, Song S, Southwell C, Swadling KM, Van de Putte A and Willis Z (2016) 'Developing priority variables ("ecosystem Essential Ocean Variables" – eEOVs) for observing dynamics and change in Southern Ocean ecosystems'. *Journal of Marine Systems*, vol 161, pp. 26–41.

Currie KI, Reid MR and Hunter KA (2011). 'Interannual variability of carbon dioxide drawdown by subantarctic surface water near New Zealand'. *Biogeochemistry*, vol 104, pp. 23–34.

Daunt F, Wanless S, Greenstreet SP, Jensen H, Hamer KC and Harris, MP (2008). 'The impact of the sandeel fishery closure on seabird food consumption, distribution, and productivity in the northwestern North Sea'. *Canadian Journal of Fisheries and Aquatic Sciences*, vol 65, no 3, pp. 362–381.

Davey PA, Pernice M, Sablok G, Larkum A, Lee HT, Golicz A, Edwards D, Dolferus R and Ralph P (2016). 'The emergence of molecular profiling and omics techniques in seagrass biology: Furthering our understanding of seagrasses'. *Functional & Integrative Genomics*, vol 16, no 5, pp. 465–480.

Davey PA, Pernice M, Ashworth J, Kuzhiumparambil U, Szabó M, Dolferus R and Ralph PJ (2017). 'A new mechanistic understanding of light-limitation in the seagrass *Zostera muelleri*'. *Marine Environmental Research*, vol 134, pp. 55–67.

Davies CH, Coughlan A, Hallegraeff G, Ajani P, Armbrecht L, Atkins N, Bonham P, Brett S, Brinkman R, Burford M, Clementson L, Coad P, Coman F, Davies D, Dela-Cruz J, Devlin M, Edgar S, Eriksen R, Furnas M, Hassler C, Hill D, Holmes M, Ingleton T, Jameson I, Leterme SC, Lønborg C, McLaughlin J, McEnnulty F, McKinnon AD, Miller M, Murray S, Nayar S, Patten R, Pritchard T, Proctor R, Purcell-Meyerink D, Raes E, Rissik D, Ruszczyk J, Slotwinski A, Swadling KM, Tattersall K, Thompson P, Thomson P, Tonks M, Trull TW, Uribe-Palomino J, Waite AM, Yauwenas R, Zammit A and Richardson AJ (2016). 'A database of marine phytoplankton abundance, biomass and species composition in Australian waters'. *Scientific Data*, vol 3, p. 160043.

Davies CH, Ajani P, Armbrecht L, Atkins N, Baird ME, Beard J, Bonham P, Burford M, Clementson L, Coad P, Crawford C, Dela-Cruz J, Doblin MA, Edgar S, Eriksen R, Everett JD, Furnas M, Harrison DP, Hassler C, Henschke N, Hoenner X, Ingleton T, Jameson I, Keesing J, Leterme SC, McLaughlin MJ, Miller M, Moffatt D, Moss A, Nayar S, Patten NL, Patten R, Pausina SA, Proctor R, Raes E, Robb M, Rothlisberg P, Saeck EA, Scanes P, Suthers IM, Swadling KM, Talbot S, Thompson P, Thomson PG, Uribe-Palomino J, van Ruth P, Waite AM, Wright S and Richardson AJ (2018). 'A database of chlorophyll *a* in Australian waters'. *Scientific Data*, vol 5, p. 180018.

Elliot JE and Elliot KH (2013). 'Tracking marine pollution'. *Science*, vol 340, no 6132, pp. 556–558, doi:10.1126/science.1235197.

Farrington JW and Takada H (2014). 'Persistent organic pollutants (POPs), polycyclic aromatic hydrocarbons (PAHs), and plastics: Examples of the status, trend, and cycling of organic chemicals of environmental concern in the ocean'. *Oceanography*, vol 27, no 1, pp. 196–213, doi:10.5670/oceanog.2014.23.

Flombaum P, Gallegos JL, Gordillo RA, Rincón J, Zabala LL, Jiao N, Karl DM, Li WKW, Lomas MW, Veneziano D, Vera CS, Vrugt JA, Martiny AC (2013). 'Present and future global distributions of the marine Cyanobacteria *Prochlorococcus* and *Synechococcus*'. *Proceedings of the National Academy of Sciences*, vol 110, pp. 9824–9829.

Furness RW (1993). 'Birds as monitors of pollutants'. In: *Birds as Monitors of Environmental Change*, RW Furness and JJD Greenwood (eds.). London: Chapman and Hall, pp. 86–143.

Furness RW (2007). 'Responses of seabirds to depletion of food fish stocks'. *Journal of Ornithology*, vol 148, no 2, pp. 247–252.

Furness RW and Camphuysen K (1997). 'Seabirds as monitors of the marine environment'. *Journal of Marine Science*, vol 54, no 4, pp. 726–737.

Furness RW, Greenwood JJD and Jarvis PJ (1993). 'Can seabirds be used to monitor the environment'. In: *Birds as Monitors of Environmental Change*, RW Furness and JJD Greenwood (eds.). London: Chapman and Hall, pp. 1–41.

Furness RW, Becker PH, Hüppop U and Davoren G (2003). 'Review of the sensitivity of seabird population in life history parameters'. In: *Seabirds as Monitors of the Marine Environment*, ML Tasker and RW Furness (eds.). ICES Cooperative Research Report, no 258, Copenhagen, pp. 26–36.

GOOS (2018). 'GOOS themes: Marine ecosystem health'. Available at: www. goosocean.org/index.php?option=com_content&view=article&id=4&Itemid=105 (accessed 12 November 2018).

Gormley AM, Slooten E, Dawson S, Barker RJ, Rayment W, DuFresne S and Brager S (2012). 'First evidence that marine protected areas can work for marine mammals'. *Journal of Applied Ecology*, vol 49, pp. 474–480.

Greig MJ, Ridgway NM and Shakespeare BS (1988). 'Sea surface temperature variations at coastal sites around New Zealand'. *New Zealand Journal of Marine and Freshwater Research*, vol 22, pp. 391–400.

Hamilton DP (2000). 'Record summer rainfall induced first recorded major cyanobacterial bloom in the Swan River'. *The Environmental Engineer*, vol 1, no 1, p. 25.

Hamilton DP, Chan T, Robb MS, Pattiaratchi CB and Herzfeld M (2001). 'The hydrology of the upper Swan River Estuary with focus on an artificial destratification trial'. *Hydrological Processes*, vol 15, no 13, pp. 2465–2480.

Hanson CE, Woo LM, Thomson PG and Pattiaratchi CB (2016). 'Observing the ocean with gliders: Techniques for data visualization and analysis'. *Oceanography*, vol 30, pp. 222–227.

Henkel LA, Nevins H, Martin M, Sugarman S, Harvey JT and Ziccardi MH (2014). 'Chronic oiling of marine birds in California by natural petroleum seeps, shipwrecks, and other sources'. *Marine Pollution Bulletin*, vol 79, no 1–2, pp. 155–163.

Hill K, Moltmann T, Proctor R and Allen S (2010). 'The Australian Integrated Marine Observing System: Delivering data streams to address national and international research priorities'. *Marine Technology Society Journal*, vol 44, pp. 65–72.

Hoegh-Guldberg O and Bruno JF (2010). 'The impact of climate change on the world's marine ecosystems'. *Science*, vol 328, pp. 1523–1528.

Hoenner X, Huveneers C, Steckenreuter A, Simpfendorfer C, Tattersall K, Jaine F, Atkins N, Babcock R, Brodie S, Burgess J, Campbell H, Heupel M, Pasquer B, Proctor R, Taylor MD, Udyawer V and Harcourt R (2018). 'Australia's continental-scale acoustic tracking database and its automated quality control process'. *Scientific Data*, vol 5, p. 170206.

Jackson JA, Carroll EL, Smith TD, Zerbini AN, Patenaude NJ and Baker CS (2016). 'An integrated approach to historical population assessment of the great whales: Case of the New Zealand southern right whale'. *Royal Society Open Science*, vol 3, p. 150669.

Jillet JB (1969). 'Seasonal hydrology of waters off the Otago Peninsula, southeastern New Zealand'. *New Zealand Journal of Marine and Freshwater Research*, vol 3, pp. 349–375.

Johnson CR, Banks SC, Barrett NS, Cazassus F, Dunstan PK, Edgar GJ, Frusher SD, Gardner C, Haddon M, Helidoniotis F, Hill KL, Holbrook NJ, Hosie GW, Last PR, Ling SD, Melbourne-Thomas J, Miller K, Pecl GT, Richardson AJ, Ridgway KR, Rintoul SR, Ritz DA, Ross DJ, Sanderson JC, Shepherd SA, Slotwinski A, Swadling KM and Taw N (2011). 'Climate change cascades: Shifts in oceanography, species' ranges and subtidal marine community dynamics in eastern Tasmania'. *Journal of Experimental Marine Biology and Ecology*, vol 400, pp. 17–32.

Kong F, Li H, Sun P, Zhou Y and Mao Y (2014). '*De novo* Assembly and characterization of the transcriptome of seagrass *Zostera marina* using illumina paired-end sequencing'. *PLoS ONE*, vol 9, p. e112245.

Lavers JL, Bond AL and Hutton I (2014). 'Plastic ingestion by Flesh-footed Shearwaters (*Puffinus carneipes*): Implications for fledgling body condition and the accumulation of plastic-derived chemicals'. *Environmental Pollution*, vol 187, pp. 124–129.

Leaper R, Cooke J, Trathan P, Reid K, Rowntree V and Payne R (2006). 'Global climate drives southern right whale population dynamics'. *Biology Letters*, vol 2, pp. 289–292.

Libro S, Kaluziak ST and Vollmer SV (2013). 'RNA-seq profiles of immune related genes in the staghorn coral *Acropora cervicornis* infected with white band disease'. *PLoS ONE*, vol 8, no 11, p. e81821.

Liu F, Wang W, Sun X, Liang Z and Wang F (2014). 'RNA-Seq revealed complex response to heat stress on transcriptomic level in *Saccharina japonica* (Laminariales, Phaeophyta)'. *Journal of Applied Phycology*, vol 26, no 3, pp. 1585–1596.

Lu Z, Zhang L, Yang J and Lin L (2009). 'The distributions of picoplankton in relation to environmental factors in Bailianjing River near Huangpu River in Shanghai, China'. Proceedings of the 2009 International Conference on Energy and Environment Technology, Guilin, Guangxi, China, 16–18 October 2009.

Lynch TP, Morello EB, Evans K, Richardson AJ, Rochester W, Steinberg CR, Roughan M, Thompson P, Middleton JF, Feng M, Sherrington R, Brando V, Tilbrook B, Ridgway K, Allen S, Doherty P, Hill K and Moltmann TC (2014). 'IMOS National Reference Stations: A continental-wide physical, chemical and biological coastal observing system'. *PLoS ONE*, vol 9, p. e113652.

McLeay LJ, Page B, Goldsworthy SD, Ward TM, Paton DC, Waterman M and Murray MD (2009). 'Demographic and morphological responses to prey depletion in a crested tern (*Sterna bergii*) population: Can fish mortality events highlight performance indicators for fisheries management?'. *ICES Journal of Marine Science*, vol 66, pp. 237–247.

McManus MA and Woodson CB (2012). 'Plankton distribution and ocean dispersal'. *Journal of Experimental Biology*, vol 215, pp. 1008–1016.

Macreadie PI, Schliep MT, Rasheed MA, Chartrand KM and Ralph PJ (2014). 'Molecular indicators of chronic seagrass stress: A new era in the management of seagrass ecosystems?'. *Ecological Indicators*, vol 38, pp. 279–281.

Macreadie PI, McLean DL, Thomson PG, Partridge JC, Jones DOB, Gates AR, Benfield MC, Collin SP, Booth DJ, Smith LL, Techera E, Skropeta D, Horton T, Pattiaratchi C, Bond T and Fowler AM (2018). 'Eyes in the sea: Unlocking the mysteries of the ocean using industrial, remotely operated vehicles (ROVs)'. *Science of The Total Environment*, vol 634, pp. 1077–1091.

Malone TC and Cole M (2000). 'Toward a global scale coastal ocean observing system'. *Oceanography*, vol 13, no 1, pp. 7–11.

Mera H and Bourne DG (2018). 'Disentangling causation: Complex roles of coral-associated microorganisms in disease'. *Environmental Microbiology*, vol 20, no 2, pp. 431–449.

Miloslavich P, Bax NJ, Simmons SE, Klein E, Appeltans W, Aburto-Oropeza O, Andersen Garcia M, Batten SD, Benedetti-Cecchi L, Checkley DM, Chiba S, Duffy JE, Dunn DC, Fischer A, Gunn J, Kudela R, Marsac F, Muller-Karger FE, Obura D and Shin YJ (2018). 'Essential ocean variables for global sustained observations of biodiversity and ecosystem changes'. *Global Change Biology*, vol 24, pp. 2416–2433.

Moore SE (2008). 'Marine mammals as ecosystem sentinels'. *Journal of Mammalogy*, vol 89, pp. 534–540.

Moreno R, Jover L, Diez C and Sanpera C (2011). 'Seabird feathers as monitors of the levels and persistence of heavy metal pollution after the Prestige oil spill'. *Environmental Pollution*, vol 159, pp. 2454–2460.

Munawar M and Weisse T (1989). 'Is the "microbial loop" an early warning indicator of anthropogenic stress?'. *Hydrobiologia*, vol 188, pp. 163–174.

Oliver ECJ, Benthuysen JA, Bindoff NL, Hobday AJ, Holbrook NJ, Mundy CN and Perkins-Kirkpatrick SE (2017). 'The unprecedented 2015/16 Tasman Sea marine heatwave'. *Nature Communications*, vol 8, p. 16,101.

Oliver ECJ, Donat MG, Burrows MT, Moore PJ, Smale DA, Alexander LV, Benthuysen JA, Feng M, Gupta AS, Hobday AJ and Holbrook NJ (2018). 'Longer and more frequent marine heatwaves over the past century'. *Nature communications*, vol 9, no 1, p. 1324.

OSPAR Commission (2008). 'Background document for the EcoQO on plastic particles in stomachs of seabirds'. Publication no 355/2008, Biodiversity Series, OSPAR Commission, London.

Partensky F, Blanchot J and Vaulot D (1999a). 'Differential distribution and ecology of *Prochlorococcus* and *Synechococcus* in oceanic waters: A review'. *Bulletin de l'Institute Oceanographique de Monaco*, vol 19, pp. 457–475.

Partensky F, Hess WR and Vaulot D (1999b). '*Prochlorococcus*, a marine photosynthetic prokaryote of global significance'. *Microbiology and Molecular Biology Reviews*, vol 63, pp. 106–127.

Pattiaratchi C, Woo LM, Thomson PG, Hong KK and Stanley D (2017). 'Ocean glider observations around Australia'. *Oceanography*, vol 30, pp. 90–91.

Pecl GT, Hobday AJ, Frusher S, Sauer WHH and Bates AE (2014). 'Ocean warming hotspots provide early warning laboratories for climate change impacts'. *Reviews in Fish Biology and Fisheries*, vol 24, pp. 409–413.

Piatt JF, Sydeman WJ and Wiese F (2007). 'Introduction: A modern role for seabirds as indicators'. *Marine Ecology Progress Series*, vol 352, pp. 199–204.

Raven JA (1998). 'The twelfth Tansley Lecture. Small is beautiful: The *picophytoplankton*'. *Functional Ecology*, vol 12, pp. 503–513.

Rees HC, Maddison BC, Middleditch DJ, Patmore JR and Gough KC (2014). 'The detection of aquatic animal species using environmental DNA – A review of eDNA as a survey tool in ecology'. *Journal of Applied Ecology*, vol 51, no 5, pp. 1450–1459.

Ridgway KR (2007). 'Long-term trend and decadal variability of the southward penetration of the East Australian Current'. *Geophysical Research Letters*, vol 34, no 13, doi:10.1029/2007GL030393.

Rolland V, Barbraud C and Weimerskirch H (2008). 'Combined effects of fisheries and climate on a migratory long-lived marine predator'. *Journal of Applied Ecology*, vol 45, no 1, pp. 4–13.

Salo T, Reusch TBH and Boström C (2015). 'Genotype-specific responses to light stress in eelgrass *Zostera marina*, a marine foundation plant'. *Marine Ecology Progress Series*, vol 519, pp. 129–140.

Sander SG, Tian F, Ibisanmi EB, Currie K, Hunter KA and Frew RD (2015). 'Spatial and seasonal variations of iron speciation in surface waters of the subantarctic front and the Otago continental shelf'. *Marine Chemistry*, vol 173, pp. 114–124.

Schmelzle MC and Kinziger AP (2016). 'Using occupancy modelling to compare environmental DNA to traditional field methods for regional-scale monitoring of an endangered aquatic species'. *Molecular Ecology Resources*, vol 16, no 4, pp. 895–908.

Shaw AGP, Kavalieris L and Vennell R (1999). 'Seasonal and interannual variability of SST off the east coast of the South Island, New Zealand'. *Geocarto International*, vol 14, pp. 29–34.

Shears NT and Bowen MM (2017). 'Half a century of coastal temperature records reveal complex warming trends in western boundary currents'. *Scientific Reports*, vol 7, p. 14527.

Seyboth E, Groch K, Dalla Rosa L, Reid K, Flores PAC and Secchi R (2016). 'Southern right whale reproductive success is influenced by krill density and climate'. *Scientific Reports* vol 6, p. 28205.

Sigsgaard EE, Nielsen IB, Bach SS, Lorenzen ED, Robinson DP, Knudsen SW, Pedersen MW, Al Jaidah M, Orlando L, Willerslev E and Møller PR (2017). 'Population characteristics of a large whale shark aggregation inferred from seawater environmental DNA'. *Nature Ecology & Evolution*, vol 1, no 1, p. 0004.

Stat M, Huggett MJ, Bernasconi R, DiBattista JD, Berry TE, Newman SJ, Harvey ES and Bunce M (2017). 'Ecosystem biomonitoring with eDNA: Metabarcoding across the tree of life in a tropical marine environment'. *Scientific Reports*, vol 7, no 1, p. 12240.

Stats NZ (2016). 'Are New Zealanders living closer to the coast?'. Available at: http://archive.stats.govt.nz/browse_for_stats/population/Migration/internal-migration/are-nzs-living-closer-to-coast.aspx (accessed 12 November 2018).

Sydeman WJ, Thompson SA, Santora JA, Koslow JA, Goericke R and Ohman MD (2015). 'Climate–ecosystem change off southern California: Time-dependent seabird predator–prey numerical responses'. *Deep Sea Research Part II: Topical Studies in Oceanography*, vol 112, pp. 158–170.

Sydnes MO (2017). 'Oil spill fingerprinting–identification of crude oil source of contamination'. In: *Petrogenic Polycyclic Aromatic Hydrocarbons in the Aquatic Environment: Analysis, Synthesis, Toxicity and Environmental Impact*, DM Pampanin and M Sydnes (eds.). Bentham Science Publishers (eBook), pp. 50–64.

Thompson PA, Baird ME, Ingleton T and Doblin MA (2009). 'Long-term changes in temperate Australian coastal waters: Implications for phytoplankton'. *Marine Ecology Progress Series*, vol 394, pp. 1–191.

Thompson PA, Bonham P, Thomson PG, Rochester W, Doblin MA, Waite AM, Richardson A and Rousseaux CS (2015). 'Climate variability drives plankton community composition changes: The 2010–2011 El Niño to La Niña transition around Australia'. *Journal of Plankton Research*, vol 37, pp. 966–984.

Thomsen PF, Møller PR, Sigsgaard EE, Knudsen SW, Jørgensen OA and Willerslev E (2016). 'Environmental DNA from seawater samples correlate with trawl catches of subarctic, deepwater fishes'. *PLoS ONE*, vol 11, no 11, p. e0165252.

Thomson PG and Pattiaratchi CB (2018). 'Trends in the abundance of picophyto-plankton due to changes in boundary currents and by marine heat waves in Australian coastal waters from IMOS National Reference Stations'. *PeerJ Preprints*, vol 6, p. e26677v1, doi:10.7287/peerj.preprints.26677v1.

Thums M, Jenner C, Waples K, Salgado Kent C and Meekan M (2018). 'Humpback whale use of the Kimberley: Understanding and monitoring spatial distribution'. Report of Project 1.2.1 prepared for the Kimberley Marine Research Program, Western Australian Marine Science Institution, Perth.

Thurber RV, Willner-Hall D, Rodriguez-Mueller B, Desnues C, Edwards RA, Angly F, Dinsdale E, Kelly L and Rohwer F (2009). 'Metagenomic analysis of stressed coral holobionts'. *Environmental Microbiology*, vol 11, pp. 2148–2163.

Twomey L and John, J (2001). 'Effects of rainfall and salt-wedge movement on phytoplankton succession in the Swan–Canning Estuary, Western Australia'. *Hydrological Processes*, vol 15, no 13, pp. 2655–2669.

Velando A, Munilla I and Leyenda PM (2005). 'Short-term indirect effects of the "Prestige" oil spill on European shags: Changes in availability of prey'. *Marine Ecology Progress Series*, vol 302, pp. 263–274.

World Bank and United Nations Department of Economic and Social Affairs (2017). 'The potential of the blue economy: Increasing long-term benefits of the sustainable use of marine resources for small island developing states and coastal least developed countries'. World Bank, Washington, DC.

Yamamoto S, Minami K, Fukaya K, Takahasi K, Sawada H, Murakami H, Tsuji S, Hashizume H, Kubonaga S, Horiuchi T, Hongo M, Nishida J, Okugawa Y, Fuji-wara A, Fukuda M, Hidaka S, Suzuki KW, Miya M, Araki H, Yamanaka H, Maruyama A, Miyashita K, Masuda R, Minamoto T and Kondoh M (2016). 'Environmental DNA as a "snapshot" of fish distribution: A case study of Japanese jack mackerel in Maizuru Bay, Sea of Japan'. *PLoS ONE*, vol 11, no 3, p. e0149786.

7 Pollution from land-based sources

*Ian Snowball, Alizée P. Lehoux,
Josie Crawshaw, Candida Savage,
Matthew R. Hipsey, Anas Ghadouani,
Sarah McCulloch and Carolyn E. Oldham*

Introduction

It is commonly asserted, following Kwiatkowski (1984), that more than 80 per cent of solid debris in the marine environment comes from land-based sources. However, as Boelens and Kershaw (2015) stated, 'there is little foundation to this assertion', because humans are biased in what they observe in coastal areas. Put simply, it is likely that much of this debris has a land-based source, but no exact figures are available, and marine pollution includes much more than objects that are visible to the naked, human eye.

For the purposes of this chapter, 'environmental pollution' is observed when one or several compound(s) in a solid, liquid or gas state are (as a result of human activity) introduced into the environment at a rate faster than they are decomposed, diluted, stored or transformed into a harmless form, and when their concentration becomes a risk to human and ecosystem health (Russell, 1974). While it is common to assume that pollutants are particles, solutes or gases, pollution can also be energy that manifests as sound, heat and radioactivity, or organisms such as pathogens. One must also appreciate that some environmental pollution commenced well before the industrial revolution. Regardless of the total pollution load from land-based sources it is important to appreciate that the relative sizes of the contributions of different types of pollution have changed as society has developed. For example, land-sourced plastic debris currently accounts for 60 to 80 per cent of total marine debris (Jambeck *et al.*, 2015), even though synthetic plastic (as 'Bakelite') was first made in 1907.

Some marine pollution has been deliberate, done when it was assumed that the ocean was such a large reservoir that anything dumped into it would be diluted to the point of becoming undetectable with no negative effect, or that solid debris (such as munitions) would fall to the ocean depths and become buried and out of harm's way for an indefinite period. Other types of pollutants, such as human-induced greenhouse gases, nutrients and a variety of solid wastes, were released irrespective of the effects they would have on the marine environment.

For some substances (and transient sound waves) there are threshold levels in either water, sediments or biota, above which their levels are considered hazardous; but there are also relatively new (emerging) pollutants for which threshold values have yet to be established. To complicate matters, there can be combined 'cocktail' (synergistic) effects of multiple pollutants that are hard to measure, and it remains difficult to assess their impact. Pollution can disrupt ecosystem functioning in many different ways, from acute toxicity of a single substance on an individual, to multiple effects on a regional scale and at many trophic levels. In some cases, waste can disrupt natural biogeochemical cycles, and completely cover benthic habitats and lead to their decline.

Some hazardous substances, such as mercury, can become even more toxic over time, while radioactive substances can decay into less harmful isotopes, and pathogens can evolve. Many organic pollutants and metals are persistent and can bioaccumulate and biomagnify in ecosystems, with their impact on marine ecosystems monitored by the health of apex predators, such as eagles and seals. Substances may also physically change, becoming smaller with the potential to be ingested by smaller organisms and then biomagnify.

This chapter explores these issues through a range of case studies from around the world, highlighting the impacts of various industries and responses to the pollution challenges. These case studies serve to demonstrate the extreme impacts human activities can have on marine and coastal environments, at a time when blue economy goals are catalysing the expansion of marine-based activities. While pollution itself is by definition an extreme impact on marine ecosystems, we also highlight how extreme events can drive risks associated with land-sourced pollutants. Lessons learnt from past experience, and current mitigation efforts, are therefore of immense value.

Industrial contaminants

As a topical case study of the effects of extreme industrial pollution on the marine environment we turn to the Baltic Sea, which is internationally recognised as one of the world's most human-impacted water bodies. All of its sub-basins and many of its coasts are considered to be unhealthy (HELCOM, 2010), largely due to the fact that it is a semi-enclosed brackish sea, partially dependent on infrequent ingressions of dense, oxygenated salt-water from the global ocean to invigorate water circulation and sustain benthic communities. Being geologically very young it hosts many species that live close to their ecological tolerances and relatively small changes in their habitat can have a large and long-lived impact on the ecosystem structure. Extended periods of stagnation and the spreading of benthic hypoxia have natural and anthropogenic drivers (Zillén et al., 2008; Kabel et al., 2012) but the recent spread of so-called dead-zones is primarily due to nutrients from land (Carstensen et al., 2014). Yet, nutrient-driven hypoxia (addressed in more detail in the next section) is not the Baltic Sea's only problem: human activities have released heavy metals, pesticides, plastics, pharmaceuticals and emerging pollutants, which all contribute to the multiple stressors that threaten the health of the Baltic Sea (HELCOM, 2010).

A manifestation of multiple contaminants entering the system was the discovery of the 'hidden-sins' in the form of 'fiberbanks' (Apler *et al.*, 2014). Fiberbanks are deposits of heavily contaminated, cellulose-rich sediments that exist at numerous locations along the Baltic Sea coast of Sweden and in some of its lakes (Norrlin and Josefsson, 2017). These fiberbanks are the legacy of a previously lucrative paper, pulp and woodboard industry that was largely ignorant of the impact it was having on the environment until awareness increased and legislation was enforced in the late 1960s. Fiberbank deposits formed when factories discharged polluted waste-water laden with solids (fibres, wood chips etc.) directly into the sea, starting in the early nineteenth century. The waste accumulated on the seafloor (or lake bed) adjacent to the factories, but was also transported further away and mixed with natural sediments to form 'fibre-rich' sediments.

Systematic inventories of fibrous sediments have only recently been completed along targeted areas of the Swedish coast. So far 28 fibrous sediment sites covering a total of $29\,km^2$ are known and, of this total, $2.5\,km^2$ consist of fiberbanks (Norrlin and Josefsson, 2017). These inventories have shown high or very high concentrations of persistent organic pollutants (POPs) and metals, such as mercury, arsenic, copper, cadmium and lead. Some of these contaminants can have chronic or acute effects on aquatic or sediment dwelling organisms. The contaminants can be transported by diffusion, gas ebullition, convection due to groundwater flow and particulate resuspension, including mass transport events (submarine landslides).

More knowledge about the physical and chemical properties of fibrous sediments is needed to develop appropriate management plans building on science-based characterisation, classification, risk assessment, prioritisation and recommendations for remediation, which are consistent with national environmental quality objectives and global sustainable development goals. The fiberbanks are situated in relatively shallow waters and will largely be uncovered due to uplift, which will expose benthic flora and fauna to the polluted sediments. The discovery of these fiberbanks has raised crucial questions that can relate to other types of contaminated sediment: What knowledge do we lack, how can we acquire it and how can we communicate it to stakeholders and environmental managers to expedite remediation?

Active remediation of fiberbanks could be done primarily in two ways: (i) the contaminated sediments can be dredged and treated somewhere else; or (ii) they can be treated *in situ* and the contamination remains enclosed in the substrate. The second solution can involve 'capping', which is normally less expensive than removal and does not require on-site treatment (Perelo, 2010). The ability of caps to chemically isolate the contaminants from benthic fauna and the physical resistance of the sediment to extra load need to be laboratory tested before pilot field tests. Various materials with different properties can be used; some active materials (such as activated carbon) have high absorption capacity and therefore can trap more pollutants, whereas regular granular material (such as crushed stones) can be used mainly to create a physical isolation layer (Jersak *et al.*, 2016). Apler *et al.* (2019) have shown that the relatively high organic carbon content of fiberbanks helps to retain metal pollutants, but it is predicted that over the course of time the organic matter will degrade through microbial activity, which will increase pore-water concentrations of metals and, in the case of mercury, cause methylation to the neurotoxic form of methylmercury.

Agricultural nutrient inputs

The enrichment of otherwise low-nutrient (oligotrophic) waters with excess nutrients is a significant issue globally. In particular, nitrogen (N) is a key nutrient for biological growth in marine systems (Gruber, 2008), with nitrate availability regulating primary production (mostly as phytoplankton) in coastal environments (Howarth, 1988; Herbert, 1999). The unprecedented increase in the human population since the late nineteenth century has dramatically altered the nitrogen cycle. The agricultural industry has expanded in response to increasing demand for food, creating fixed nitrogen fertilisers that increase land productivity (Tilman, 1999). The runoff of excess fertilisers can have detrimental effects on downstream aquatic ecosystems, including rivers, lakes, lowland lagoons and estuaries (McDowell and Wilcock, 2008). In combination with emissions of N to the atmosphere caused by the burning of fossil fuels (Vitousek *et al.*, 1997; McCrackin and Elser, 2010), anthropogenic N production has increased by a factor of ten since the late nineteenth century (Galloway *et al.*, 2004).

Eutrophication can be defined as 'an increase in the rate of supply of organic matter to an ecosystem' (Nixon, 2012). Worldwide estuarine health is reported to be in decline, exhibiting one or more eutrophication symptoms due to intensified agricultural activities, urban runoff, wastewater treatment plants and atmospheric N deposition (Bricker *et al.*, 2008). Two key drivers of change in estuaries are extreme loads of nutrients and sediments. Increased inputs of clays and silts from land clearance smother benthic invertebrates and plants (Thrush *et al.*, 2013) and reduce water clarity (Cantilli *et al.*, 2006), which negatively impacts primary producers (Cloern, 2001). Increased nutrients in coastal ecosystems elevate phytoplankton biomass, which can have deleterious effects, such as increasing light attenuation in the waters and limiting the growth of rooted macrophytes, including seagrass (Cloern, 2001; Seitzinger *et al.*, 2006; Bargu *et al.*, 2011). Benthic microalgae can also be shaded, decreasing gross primary production (Meyercordt *et al.*, 1999; Krause-Jensen *et al.*, 2012). The resulting shift from perennial macroalgae and seagrass dominated communities or benthic microalgae towards communities dominated by ephemeral macroalgae and pelagic microalgae alters habitat quality for animals (Cloern, 2001; Borum, 2013). Shifts in the community composition and relative abundance of macroinvertebrates also occurs with increasing eutrophication pressure (Pearson and Rosenberg, 1978; Bužančić *et al.*, 2016). As phytoplankton blooms die and decompose, oxygen is consumed rapidly from the water column by microorganisms, and hypoxia can occur in bottom waters and in sediment (Diaz and Rosenberg, 2008).

As a case study, we look at coastal eutrophication issues in New Zealand, where a 60-fold increase in the use of nitrogenous fertilisers occurred between 1961 and 2001, in line with a steady increase in irrigation and intensification of farming practices (MacLeod and Moller, 2006). New Zealand pastoral streams are often rich in nitrates and algal biomass compared to native forest streams (Quinn *et al.*, 1997; McDowell and Wilcock, 2008) and water quality is very poor in lowland streams in pastoral catchments.

The consequences of such substantial land-use changes are exemplified based on the New Zealand Southland region where land use shifted from sheep to dairy farming at the same time as rainfall increased in the region and, consequently, there were further increased nutrient and sediment inputs to estuarine ecosystems. The New River Estuary is a large (41 km^2 at high tide) tidal estuary near Invercargill city. Eight catchments drain into the estuary including two large rivers (Oreti and Waihopai) that pass through dairy farming (Cavanagh and Ward, 2014). Sedimentation rates in New River Estuary have increased from 13–16 mm per year during the period 1967–2007, up to rates of 72 mm per year in 2011 (Robertson and Stevens, 2013). Nuisance macroalgae have expanded from <1 per cent in 2001 to occupying 8 per cent of the estuary by 2009, likely driven by the high nitrogen loads from the catchment (estimated N ~100 mg m^{-2} day) (Robertson and Stevens, 2011). The increased nuisance algae trap fine muds, create anoxic sediment conditions and produce sulphides (H$_2$S) (Cloern, 2001). Decomposing algae will release bioavailable nitrogen back into the water column, predominantly as ammonium, supporting continued growth of nuisance algae. Macrofauna are negatively affected, with shifts in abundance of the community to a dominance of tolerant surface-feeding taxa that can withstand the anoxic sediment conditions (Robertson and Stevens, 2013). The potential negative consequences of eutrophication described here are however dependent on how quickly the nutrients are distributed and removed from the system by flushing or microbial transformations (Pinckney *et al.*, 2001), which may buffer nitrogen pollution in estuaries.

One of the primary nitrogen removal pathways in coastal and estuarine environments is denitrification (Pinckney *et al.*, 2001). Denitrification is a microbial process predominantly carried out at the oxic/anoxic interface of sediments (Herbert, 1999) where denitrifying microorganisms convert nitrate (NO$_3^-$) to biologically unavailable gaseous compounds such as dinitrogen gas (N$_2$) and the greenhouse gas, nitrous oxide (N$_2$O) (Herbert, 1999; Fennel *et al.*, 2009). Accordingly, the process of denitrification is critical in removing excess anthropogenic inputs of nitrogen from coastal environments (Piña-Ochoa and Álvarez-Cobelas, 2006; Seitzinger *et al.*, 2006). Hotspots of denitrification may occur where there are higher concentrations of organic matter and the presence of bioturbating macrofauna (Crawshaw *et al.*, 2018). These sedimentary zones may be important habitats in supporting natural remediation of nitrogen pollution in coastal ecosystems. While we have a good understanding of how nitrogen loading, and concomitant effects such as hypoxia, affect denitrification (e.g. Gongol and Savage, 2016; Douglas *et al.*, 2017), the indirect effects of the multiple stressors of sedimentation and nitrogen loading on nitrogen cycling requires further investigation.

Urbanisation

While agriculture is the most significant driver of nutrients to the coast, the rapid rate of urbanisation experienced around estuaries, deltas and coastal embayments has also had a considerable impact worldwide. Cities predominantly contribute nutrients to the coast through stormwater and waste-water discharges, which can be greatly amplified during extreme floods that drive urban runoff and overflows of sewers. Of increasing interest in the research community is also the potential contribution of Submarine Groundwater Discharge (SGD), whereby nutrient polluted groundwater enters the coastal margin.

As a case study of the complexities of urban nutrient input we turn to the Swan Coastal Plain in Western Australia. Urbanisation of the Perth region has increased rapidly since the 1980s, and over the last decade has extended into what was previously drained agricultural areas (Ocampo and Oldham, 2017). Projected future urban development continues to move into water-logged areas that require drainage to expedite water discharge to the ocean. Simultaneously, there has been increased focus on the environmental protection of the oligotrophic coastal waters in this region, which are susceptible to nutrient inputs from urban drainage (e.g., Cottingham *et al.*, 2018). As a result, environmental targets for drainage water are orders of magnitude lower than found in Europe and the USA, making management of urban discharge challenging.

Coincident with the need to make large reductions in nutrient discharges, the sands of the Swan Coastal Plain are highly permeable, with high recharge rates, a relatively shallow groundwater table and event-driven interflows (Ocampo and Oldham, 2017). A century of clearing deep-rooted perennial native vegetation and replacement with shallow-rooted vegetation or impervious surfaces has significantly perturbed the water balance (Davies *et al.*, 2017). Use of the increasingly water-logged areas has required extensive drainage of the seasonally high water tables, and frequently these drains discharge directly to local wetlands or creeks, ultimately carrying nutrients to the coastal plain rivers and estuaries.

To control nutrient inputs there is now an increased interest in water sensitive urban design (WSUD) technologies, which discourage direct discharge of drainage water to local streams and wetlands and encourage increased local infiltration (Ocampo and Oldham, 2017). The latter may be used to manage subsequent re-use of drainage water (e.g. irrigation of public open space) and also to ensure a sub-surface pathway and therefore promote nutrient attenuation. In areas with high groundwater, this creates a conflict between rehabilitation of urban drains into 'living streams' or infiltration basins, drainage to prevent potential groundwater flooding and the protection of receiving waters from unwanted nutrient loads. The high groundwater creates complexities in determining the water balance and nutrient attenuation performance of WSUD features that receive seasonal inputs of groundwater and its nutrient load (Ocampo, 2018).

Recent research has highlighted the benefits of the sub-surface pathway for nutrient attenuation of stormwater (Adyel et al., 2016). Such novel stormwater treatment wetlands can employ multiple flow compartments within a parkland context to improve the urban landscape while also reducing nutrient loads to coastal waters, even under a large range of hydrologic conditions spanning droughts to floods.

This case study highlights the challenges of protecting oligotrophic coastal waters that lie adjacent to urban centres situated on sandy coastal plains and experiencing high groundwater. Historically, this groundwater has drained towards the coast via the sub-surface pathway, yet urbanisation has brought these waters to the surface and significantly reduced travel times and nutrient attenuation, resulting in the rapid transport of nutrient-rich drainage waters. By designing stormwater treatment wetlands using alternating surface and sub-surface pathways, nutrient attenuation can be achieved prior to discharge to receiving waters and ultimately to the coastal zone. It provides a useful example of current management approaches designed to respond to growing urban challenges and mitigate past land-use changes.

Microbial pollution and pathogens

Of particular concern in coastal environments are the potential risks created by pathogens (Belkin and Colwell, 2006). Microbial pollution is, however, broader than pathogens alone and also refers to microbial indicator organisms (usually Faecal Indicator Organisms, FIOs) or other microbial contaminants that are not naturally present in high abundance. In most cases, our management focus is on pathogens that are a human health risk. These are typically organisms that would circulate within humans, but somehow enter the marine environment. Some indigenous marine organisms can, however, also create a challenge for managers (e.g. *Vibrio* spp.). Non-human pathogens must also be considered, for example, the introduction of pests that impact local organisms of high ecological or economic value. Together, microbial pollution is therefore a complex mix of protozoan, bacterial and viral organisms that present a public health risk.

To understand and manage microbial pollution in any given coastal setting we must: (a) understand pathways of input; (b) understand biophysical processes within the water impacting organism viability; and (c) assess routes of exposure. The nature of inputs, routes of exposure and processes controlling fate are all a function of spatio–temporal variability in coastal processes (as summarised in Chapter 3), land-use activities in the surrounding catchments (e.g. residential, industrial, agricultural) and socio-economic uses of the coast (e.g. recreation, transport, aquaculture).

Like the nutrients addressed in the previous section, major microbial and pathogen pollution events and subsequent exposure risks follow extremes associated with hydro-meteorological phenomena, such as flood waters in coastal catchments that mobilise pathogens and indicator organisms into rivers, stormwater systems and eventually estuaries and the wider coastal margin. Agricultural catchments that discharge to the sea are notoriously high in pathogen loads, but this is highly dependent on specific land-use types, animal husbandry practices and other mitigating environmental conditions (Hipsey and Brookes, 2013). Urban systems have a different source profile (Jovanovic *et al.*, 2015). Despite sophisticated sewer treatment systems in most developed nations, runoff from urban catchments can still be high in wash-off from animal faeces, and urban stormwater systems are known to be a large source (Ahn *et al.*, 2005), even under dry flow conditions (Rippy *et al.*, 2014). Of more significance are floods that overload the hydraulic capacity of wastewater treatment plants (WWTP), and cause overflows to downstream receiving waters; these events cause sewage to bypass treatment systems and often lead to high loads of viable pathogen and microbial indicator organisms entering the coast as a discrete pulse (Al Aukidy and Verlicchi, 2017). Direct sewage input in developing nations also remains a challenge. Aside from river or sewer inputs, sources can also come directly from human uses, for example, loads are known to be high in beaches used heavily by recreation and often coincide with hot days or social events (Graczyk *et al.*, 2007; Abdelzaher *et al.*, 2010). Other coastal activities associated with ports, harbours and aquaculture facilities (see Chapter 10) are also known to lead to loads entering the water column.

Once organisms have entered the coast, they experience a myriad of potentially hostile environmental conditions. These include temperatures, salinities and pH values that are generally far from their optimal environment, in addition to exposure to sunlight and predation pressures (Hipsey *et al.*, 2008). Sensitivities to these conditions are, however, highly specific to different organism classes, and can even vary widely among species that are otherwise quite similar. Where organisms are able to attach to suspended sediment material, they are known to settle and can accumulate in the sediment pool, which can become self-sustaining under the right conditions (Lee *et al.*, 2006). Deposited organisms are known to be remobilised into high risk zones during extreme wind-wave conditions or under conditions of high currents (Le Fevre and Lewis, 2003; Roslev *et al.*, 2008).

Monitoring microbial pollution is more challenging than for other physico-chemical pollutants. Issues associated with detection limits, viable but non-culturable status of pathogenic organisms and sampling difficulties during extreme events have meant that often agencies rely on FIOs for assessing risk (Schang *et al.*, 2016), or other 'easy to measure' surrogate indicators (e.g. caffeine, turbidity). Hydrodynamic drivers of transport and resuspension can explain to a large extent spatio–temporal variability in both pathogens and surrogates (Lim *et al.*, 2017; McCarthy *et al.*, 2017; Eregno *et al.*, 2018), though a reliance on surrogate indicators for assessing true pathogen risk can be problematic, leading to both over- and under-prediction of the actual risk (e.g. Grant *et al.*, 2001; Cantwell *et al.*, 2016).

The risks to human health of microbial pollution are highly site and context specific, and coastal managers need to design risk assessments accordingly (Hipsey and Brookes, 2013; Lim *et al.*, 2017). Generally, direct contact following recreation or consumption of infected shellfish is the primary route of exposure in coastal systems. Established frameworks for quantitative microbial risk assessment (QMRA) are recommended to identify the varied pathways of source and exposure (Ashbolt *et al.*, 2010). Outcomes from such assessment lead to the prioritisation of options to manage the risk, which include beach closures, restrictions in seafood harvest and managing the inputs at source, through constructed wetlands, for example.

Microplastics and microfibres

Among the other pollution challenges addressed above plastic pollution is now acknowledged as one of the biggest threats to marine ecosystem futures (Haward, 2018). This global problem is considered or projected to achieve the status of a planetary boundary, which will result in unavoidable and potentially irreversible disruption to ecosystem functions (Villarrubia-Gómez *et al.*, 2018). It is especially concerning because plastic pollution has reached a global scale before being recognised as a significant or serious threat to the environment, and it almost took authorities worldwide by surprise. While there is no doubt that the problem is now acknowledged at local, regional, national and global political levels, there are still many levels of confusion about the sources and especially the pathways of plastic pollutants, in comparison to other persistent and toxic chemicals (Browne, 2015). The qualification of plastic pollution as a toxic and persistent chemical pollutant is only gradually being accepted by environmental authorities, and the magnitude and serious nature of the problem has only just been appreciated. In addition to the volume of plastic litter, the problem could be significantly greater when smaller particles, known as microplastics (<5 mm) and microfibres, are considered. The sources, the pathways and the fates of this subclass of plastic pollution, known as 'micro-litter', are potentially even more complex to investigate and manage given the current lack of information (Magnusson and Norén, 2014).

In recent years WWTPs have been identified as major point sources for the pollution of microplastics into waterways and recent studies show that the loads of microplastics to rivers, lakes, oceans and shorelines is increasing (Magnusson and Norén, 2014; Dris *et al.*, 2015). The negative effects of this microplastic contamination on aquatic environments, particularly aquatic organisms, are well-documented, with severe health and environmental implications due to the entanglement, gut-blockage or pseudo-satiation of aquatic fauna. This concern is heightened by the understanding that 56 per cent of polymers consist of level IV or level V monomers, the second highest and highest ranking environmental and health hazards, respectively (Lithner *et al.*, 2011; Miranda and de Carvalho-Souza, 2016). Therefore, developing an approach to limit or reduce contamination by microplastics as a direct result of WWTPs is essential and of the highest priority.

Since WWTPs have been identified as significant point sources for pollution, recent studies have aimed to measure the precise level of microplastic contamination as a direct result of WWTPs (Mason *et al.*, 2016; Murphy *et al.*, 2016; Ziajahromi *et al.*, 2016; Kalcikova *et al.*, 2017; Leslie *et al.*, 2017; Duran *et al.*, 2018). The definition of microplastics is often flexible; however, for measurement in waste-water, microplastics are most widely defined as plastics with the dimensions of less than 5 mm in each measurement (Tagg *et al.*, 2015). It is understood that microplastics are entering the waste-water system as either primary or secondary microplastics; these originate from either microbeads that are designed for facewashes and toothpastes or as microfibres that have decomposed from fibres used to make synthetic textiles and entered the waste-water system when washed or, alternatively, as secondary microplastics produced by decomposing macroplastic materials (Barnes *et al.*, 2009; Browne *et al.*, 2010; Browne *et al.*, 2011).

A formal, standardised methodology for the identification and classification of microplastics in waste-water is yet to be collectively agreed upon across the scientific community. This drawback has resulted in significant discrepancies across the results from recent studies (Ziajahromi *et al.*, 2016). From a sample of seven studies carried out in WWTPs across Finland, France, Scotland, USA, Sweden, Russia and Australia, the number of plastic particles per litre of waste-water found to be entering the WWTP ranges from 1 particle/L (Carr *et al.*, 2016) to 610 particles/L (Talvitie *et al.*, 2015) and the number of particles found to be leaving the system ranged from 7·10–4 particles/L (Carr *et al.*, 2016) to 50 particles/L (Dris *et al.*, 2015). As a result of these discrepancies Tagg *et al.* (2015) aimed to create a standardised methodology for the identification and quantification of microplastics in waste-water by designing and validating a suitable methodology. This process was found to have a 98.33 per cent success rate for accurate microplastic detection of particles with a length greater than 150 μm (Tagg *et al.*, 2015). The discrepancies in the results pertaining to the characterisation and quantification of microplastics in WWTPs across studies is also an illustration of how little is known about WWTPs as a point source for microplastic contamination. Urban runoff through the stormwater drainage system following rainfall events or the application of municipal sludge on agricultural lands are two other potential pathways that could contribute significantly to the transport of microplastics and microfibres to marine systems.

Reducing microplastics at their source could be an effective management strategy to reduce the release of particles into the environment. But without a better understanding of the flow of microplastics from land to the oceans, and their subsequent accumulation in the marine environment, effective management methods cannot be decided upon and prioritised.

Conclusion

This chapter has explored a range of case studies to highlight the extreme impacts of industrial and agricultural activities over time as well as those of increasing urbanisation. In some cases, efforts are still being made to address legacy pollution issues. In other situations, novel approaches have been identified to respond to growing urban pressures. In all cases the consequences of human activities have been shown to be extreme unless well-planned and early mitigation efforts are implemented. There are clearly lessons to be learnt in the context of blue economy developments, which offer many benefits but come with the risk of impacting upon marine and coastal environmental health and ultimately human health.

References

Abdelzaher, A. M., Wright, M. E., Ortega, C., Solo-Gabriele, H. M., Miller, G., Elmir, S., Newman, X., Shih, P., Bonilla, J. A., Bonilla, T. D., Palmer, C. J., Scott, T., Lukasik, J., Harwood, V. J., McQuaig, S., Sinigalliano, C., Gidley, M., Plano, L. R. W., Zhu, X., Wang, J. D. and Fleming, L. E. (2010). 'Presence of pathogens and indicator microbes at a non-point source subtropical recreational marine beach'. *Applied and Environmental Microbiology*, vol 76, no 3, pp. 724–732.

Adyel, T., Oldham, C. and Hipsey. M. (2016). 'Stormwater nutrient attenuation in a constructed wetland with alternating surface and subsurface flow pathways: Event to annual dynamics'. *Water Research*, vol 107, pp. 66–82.

Ahn J. H., Grant S. B., Surbeck C. Q., DiGiacomo P. M., Nezlin N. P., Jiang S (2005). 'Coastal water quality impact of stormwater runoff from an urban watershed in southern California'. *Environmental Science & Technology*, vol 39, no 16, pp. 5940–5953.

Al Aukidy, M. and Verlicchi, P. (2017). 'Contributions of combined sewer overflows and treated effluents to the bacterial load released into a coastal area'. *Science of the Total Environment*, vol 607, pp. 483–496.

Apler, A., Nyberg, J., Jönsson, K., Hedlund, I., Heinemo, S-Å and Kjellin, B. (2014). 'Kartläggning av fiberhaltiga sediment längs Väternorrlands kust'. SGU–rapport 2014:16, [Geological Survey of Sweden report no 2014:16; in Swedish with a summary in English], Uppsala, Sweden.

Apler, A., Snowball, I., Frogner-Kockum, P. and Josefsson, S. (2019). 'Distribution and dispersal of metals in contaminated fibrous sediment of industrial origin'. *Chemosphere*, vol 219, pp. 470–481.

Ashbolt, N. J., Schoen, M. E., Soller, J. A. and Roser, D. J. (2010). 'Predicting pathogen risks to aid beach management: The real value of quantitative microbial risk assessment (QMRA)'. *Water Research*, vol 44, no 16, pp. 4692–4703.

Bargu, S., White. J., Li, C., Czubakowski, J. and Fulweiler, R. (2011). 'Effects of freshwater input on nutrient loading, phytoplankton biomass, and cyanotoxin production in an oligohaline estuarine lake'. *Hydrobiologia*, vol 661, pp. 377–389.

Barnes, D. K. A., Galgani, F., Thompson, R. C. and Barlaz, M. (2009). 'Accumulation and fragmentation of plastic debris in global environments'. *Philosophical Transactions of the Royal Society B – Biological Sciences*, vol 364, pp. 1985–1998.

Belkin, S. and Colwell, R. R. (eds.) (2006). *Oceans and Health: Pathogens in the Marine Environment*. New York: Springer.

Boelens, R. and Kershaw, P. J. (eds.) (2015). 'Pollution in the open oceans 2009–2013 – A report by a GESAMP Task Team'. GESAMP [Joint Group of Experts on the Scientific Aspects of Marine Environmental Pollution] Reports and Studies no 91, London.

Borum, J. (2013). 'Shallow waters and land/sea boundaries', in B. B. Jørgensen and K. Richardson (eds.), *Eutrophication in Coastal Marine Ecosystems*. Washington, DC: American Geophysical Union, pp. 179–203.

Bricker, S. B., Longstaff, B., Dennison, W., Jones, A., Boicourt, K., Wicks, C. and Woerner, J. (2008). 'Effects of nutrient enrichment in the nation's estuaries: A decade of change'. *Harmful Algae*, vol 8, pp. 21–32.

Browne, M. A. (2015). *Sources and Pathways of Microplastics to Habitats*. Berlin: Springer-Verlag Berlin.

Browne, M. A., Galloway, T. S. and Thompson, R. C. (2010). 'Spatial patterns of plastic debris along estuarine shorelines'. *Environmental Science & Technology*, vol 44, no 9, pp. 3404–3409.

Browne, M. A., Crump, P., Niven, S. J., Teuten, E., Tonkin, A., Galloway, T. and Thompson, R. (2011). 'Accumulation of microplastic on shorelines worldwide: Sources and sinks'. *Environmental Science & Technology*, vol 45, no 21, pp. 9175–9179.

Bužančić, M., Ninčević Gladan, Ž., Marasović, I., Kušpilić, G. and Grbec, B. (2016). 'Eutrophication influence on phytoplankton community composition in three bays on the eastern Adriatic coast', *Oceanologia*, vol 58, pp. 302–316.

Cantilli, R., Stevens, R., Sweitlik, W., Berry, W., Kaufman, P., Paul, J., Spehar, R., Cormier, S. and Norton, D. (2006). 'Framework for developing suspended and bedded-sediments (SABS): Water quality criteria'. US Environmental Protection Agency, Office of Water, Washington, DC.

Cantwell, M. G., Katz, D. R., Sullivan, J. C., Borci, T. and Chen, R. F. (2016). 'Caffeine in Boston Harbor past and present, assessing its utility as a tracer of wastewater contamination in an urban estuary'. *Marine Pollution Bulletin*, vol 108 no 1–2, pp. 321–324. doi:10.1016/j.marpolbul.2016.04.006.

Carr, S. A., Liu, J. and Tesoro, A. G. (2016). 'Transport and fate of microplastic particles in wastewater treatment plants'. *Water Research*, vol 91, pp. 174–182.

Carstensen, J., Andersen, J. H., Gustafsson, B. G. and Conley, D. J. (2014). 'Deoxygenation of the Baltic Sea during the last century'. *Proceedings of the National Academy of Sciences*, vol 111, pp. 5628–5633.

Cavanagh, J. E. and Ward, N. (2014). 'Contaminants in estuarine and riverine sediments and biota in Southland'. Environment Southland report, Invercargill, New Zealand.

Cloern, J. E. (2001). 'Our evolving conceptual model of the coastal eutrophication problem'. *Marine Ecology Progress Series*, vol 210, pp. 223–253.

Cottingham, A., Huang, P., Hipsey, M. R., Hall, N. G., Ashworth, E., Williams, J. and Potter, I. C. (2018). 'Growth, condition, and maturity schedules of an estuarine fish species change in estuaries following increased hypoxia due to climate change'. *Ecology and Evolution*, vol 8, pp. 7111–7130.

Crawshaw, J., Schallenberg, S. and Savage, C. (2018). 'Physical and biological drivers of sediment oxygenation and denitrification in a New Zealand intermittently closed and open lake lagoon'. *New Zealand Journal of Marine and Freshwater Research*. In press. doi:10.1080/00288330.2018.1476388.

Davies, C. G, Vogwill. R. and Oldham, C. (2017). 'Urban sub-surface drainage as an alternative water source in a drying climate'. *Australasian Journal of Water Resources*, vol 20, no 2, pp. 148–159.

Diaz, R. J. and Rosenberg, R. (2008). 'Spreading dead zones and consequences for marine ecosystems'. *Science*, vol 321, no 5891, pp. 926–929.

Douglas, E. J., Pilditch, C. A., Kraan, C., Schipper, L. A., Lohrer, A. M. and Thrush, S. F. (2017). 'Macrofaunal functional diversity provides resilience to nutrient enrichment in coastal sediments'. *Ecosystems*, vol 20, pp. 1324–1336.

Dris, R., Imhof, H., Sanchez, W., Gasperi, J., Galgani, F., Tassin, B. and Laforsch, C. (2015). 'Beyond the ocean: Contamination of freshwater ecosystems with (micro-) plastic particles'. *Environmental Chemistry*, vol 12, no 5, pp. 539–550.

Duran, F. E., de Araujo, D. M., Brito, C. D., Santos, E. V., Ganiyu, S. O. and Martinez-Huitle, C. A. (2018). 'Electrochemical technology for the treatment of real washing machine effluent at pre-pilot plant scale by using active and non-active anodes'. *Journal of Electroanalytical Chemistry*, vol 818, pp. 216–222.

Eregno, F. E., Tryland, I., Tjomsland, T., Kempa, M. and Heistad, A. (2018). 'Hydrodynamic modelling of recreational water quality using *Escherichia coli* as an indicator of microbial contamination'. *Journal of Hydrology*, vol 561, pp. 179–186.

Fennel, K., Brady, D., DiToro. D., Fulweiler, R. W., Gardner, W. S., Giblin, A., McCarthy, M. J., Rao, A., Seitzinger, S., Thouvenot-Karppoo, M. and Tobias, C. (2009). 'Modelling denitrification in aquatic sediments'. *Biogeochemistry*, vol 93, pp. 159–178.

Galloway, J. N., Dentener, F. J., Capone, D. G., Boyer, E. W., Howarth, R. W., Seitzinger, S. P., Asner, G. P., Cleveland, C. C., Green, P. A., Holland, E. A., Karl, D. M., Michaels, A. F., Porter, J. H., Townsend, A. R. and Vöosmarty, C. J. (2004). 'Nitrogen cycles: Past, present, and future'. *Biogeochemistry*, vol 70, pp. 153–226.

Gongol, C. and Savage, S. (2016). 'Spatial variation in rates of benthic denitrification and environmental controls in four New Zealand estuaries'. *Marine Ecology Progress Series*, vol 556, pp. 59–77.

Graczyk, T. K., Sunderland, D., Tamang, L., Lucy, F. E. and Breysse, P. N. (2007). 'Bather density and levels of *Cryptosporidium*, *Giardia*, and pathogenic microsporidian spores in recreational bathing water'. *Parasitology Research*, vol 101, no 6, pp. 1729–1731. doi:10.1007/s00436-007-0734-1.

Grant, S. B., Sanders, B. F., Boehm, A. B., Redman, J. A., Kim, J. H., Mrše, R. D., Chu, A. K., Gouldin, M., McGee, C. D., Gardiner, N. A., Jones, B. H., Svejkovsky, J., Leipzig, G. V. and Brown A. (2001). 'Generation of Enterococci bacteria in a coastal saltwater marsh and its impact on surf zone water quality'. *Environmental Science & Technology*, vol 35, no 12, pp. 2407–2416. doi:10.1021/es0018163.

Gruber, N. (2008). 'The marine nitrogen cycle: Overview and challenges', in D. G. Capone, D. A. Bronk, M. R. Mulholland and E. J. Carpenter (eds.), *Nitrogen in the Marine Environment* (2nd edn.). San Diego, CA: Academic Press, pp. 1–50.

Haward, M. (2018). 'Plastic pollution of the world's seas and oceans as a contemporary challenge in ocean governance'. *Nature Communications*, vol 9, pp. 1–3. doi:10.1038/s41467-018-03104-3.

HELCOM (2010). 'Ecosystem health of the Baltic Sea'. Baltic Sea Environment Proceedings, no 122, Helsinki Commission, Helsinki.

Herbert, R. A. (1999). 'Nitrogen cycling in coastal marine ecosystems'. *FEMS Microbiology Reviews*, vol 23, pp. 563–590.

Hipsey, M. R. and Brookes, J. D. (2013). 'Pathogen management in surface waters: Practical considerations for reducing public health risk', in A. J. Rodriguez-Morales. (ed.), *Current Topics in Public Health*. Rijeka: IntechOpen, pp. 446–475.

Hipsey, M. R., Antenucci, J. P. and Brookes, J. D. (2008). 'A generic, process-based model of microbial pollution in aquatic systems'. *Water Resources Research*, vol 44, no 7, p. W07408. doi:10.1029/2007WR006395.

Howarth, R. W. (1988). 'Nutrient limitation of net primary production in marine ecosystems'. *Annual Review of Ecology and Systematics*, vol 19, pp. 89–110.

Jambeck, J. R., Geyer, R., Wilcox, C., Siegler, T. R., Perryman, M., Andrady, A., Narayan, R. and Law, K. L. (2015). 'Plastic waste inputs from land into the ocean'. *Science*, vol 347, no 6223, pp. 768–771.

Jersak, J., Göransson, G., Ohlsson, Y., Larsson, L., Flyhammar, P. and Lindh, P. (2016). 'In-situ capping of contaminated sediments. In-situ capping of Sweden´s fiberbank sediments: A unique challenge'. SGI Publication 30–5E, Swedish Geotechnical Institute, Linköping.

Jovanovic, D., Henry, R., Coleman, R., Deletic, A. and McCarthy, D. (2015). 'Integrated conceptual modelling of faecal contamination in an urban estuary catchment'. *Water Science & Technology*, vol 72, no 9, pp. 1472–1480. doi:10.2166/wst.2015.363.

Kabel, K., Moros, M., Porsche, C., Neumann, T., Adolphi, F., Anderson, T. J., Siegel, H., Gerth, M., Leipe, T., Jansen, E. and Sinnighe Damsté, J. S. (2012). 'Impact of climate change on the Baltic Sea ecosystem over the past 1,000 years'. *Nature Climate Change*, vol 2, pp. 871–874.

Kalcikova, G., Alic, B., Skalar, T., Bundschuh, M. and Gotvajn, A. Z. (2017). 'Wastewater treatment plant effluents as source of cosmetic polyethylene microbeads to freshwater'. *Chemosphere*, vol 188, pp. 25–31.

Krause-Jensen, D., Markager, S. and Dalsgaard, T. (2012). 'Benthic and pelagic primary production in different nutrient regimes'. *Estuaries and Coasts*, vol 35, pp. 527–545.

Kwiatkowski, B. (1984). Marine pollution from land-based sources: Current problems and prospects. *Ocean Development and International Law*, vol 14, pp. 315–335.

Lee, C. M., Lin, T. Y., Lin, C-C., Kohbodi, G. A., Bhatt, A., Lee, R. and Jay, J. A. (2006). 'Persistence of fecal indicator bacteria in Santa Monica Bay beach sediments'. *Water Research*, vol 40, pp. 2593–2602.

Le Fevre, N. M. and Lewis, G. D. (2003). 'The role of resuspension in enterococci distribution in water at an urban beach', *Water Science & Technology*, vol 47, no 3, pp. 205–210.

Leslie, H. A., Brandsma, S. H., van Velzen, M. J. M. and Vethaak, A. D. (2017). 'Microplastics en route: Field measurements in the Dutch river delta and Amsterdam canals, wastewater treatment plants, North Sea sediments and biota'. *Environment International*, vol 101, pp. 133–142.

Lim, K. Y., Shao, S., Peng, J., Grant, S. B. and Jiang, S. C. (2017). 'Evaluation of the dry and wet weather recreational health risks in a semi-enclosed marine embayment in Southern California'. *Water Research*, vol 111, pp. 318–329.

Lithner, D., Larsson, A. and Dave, G. (2011). 'Environmental and health hazard ranking and assessment of plastic polymers based on chemical composition'. *Science of the Total Environment*, vol 409, no 18, pp. 3309–3324.

McCarthy, D. T., Jovanovic, D., Lintern, A., Teakle, I., Barnes, M., Deletic, A. and Hipsey, M. R. (2017). 'Source tracking using microbial community fingerprints: Method comparison with hydrodynamic modelling'. *Water Research*, vol 109, pp. 253–265.

McCrackin, M. L. and Elser, J. J. (2010). 'Atmospheric nitrogen deposition influences denitrification and nitrous oxide production in lakes'. *Ecology*, vol 91, pp. 528–539.

McDowell, R. W. and Wilcock, R. J. (2008). 'Water quality and the effects of different pastoral animals'. *New Zealand Veterinary Journal*, vol 56, pp. 289–296.

MacLeod, C. J. and Moller, H. (2006). 'Intensification and diversification of New Zealand agriculture since 1960: An evaluation of current indicators of land use change'. *Agriculture, Ecosystems & Environment*, vol 115, pp. 201–218.

Magnusson, K. and Norén, F. (2014). 'Screening of microplastic particles in and down-stream a wastewater treatment plant'. IVL Swedish Environmental Institute report, no C55, Stockholm.

Mason, S. A., Garneau, D., Sutton, R., Chu, Y., Ehmann, K., Barnes, J., Fink, P., Papazissimos, D. and Rogers, D. L. (2016). 'Microplastic pollution is widely detected in US municipal wastewater treatment plant effluent'. *Environmental Pollution*, vol 218, pp. 1045–1054.

Meyercordt, J., Gerbersdorf, S. and Meyer-Reil, L-A. (1999). 'Significance of pelagic and benthic primary production in two shallow coastal lagoons of different degrees of eutrophication in the southern Baltic Sea'. *Aquatic Microbial Ecology*, vol 20, pp. 273–284.

Miranda, D. D. and de Carvalho-Souza, G. F. (2016). 'Are we eating plastic-ingesting fish?'. *Marine Pollution Bulletin*, vol 103, no 1–2, pp. 109–114.

Murphy, F., Ewins, C., Carbonnier, F. and Quinn, B. (2016). 'Wastewater treatment works (WwTW) as a source of microplastics in the aquatic environment'. *Environmental Science & Technology*, vol 50, no 11, pp. 5800–5808.

Nixon, S. W. (2012). 'Coastal marine eutrophication: A definition, social causes, and future concerns'. *Ophelia*, vol 41, no 1, pp. 199–219.

Norrlin, K. and Josefsson, S. (2017). 'Förorenad fibersediment i svenska hav och sjöar'. SGU-rapport 2017:07, [Geological Survey of Sweden report no 2017:07; in Swedish with a summary in English], Uppsala, Sweden.

Ocampo, C. (2018). 'The impact of urbanization and stormwater management practices on water balances and nutrient pathways in areas of high groundwater: A review of recent literature'. CRC Water Sensitive Cities report, Melbourne.

Ocampo C, and Oldham C. (2017). 'Hydrologic performance of a bioretention basin affected by groundwater intrusion'. *Proceedings of the 37th International Association of Hydraulic Engineering and Research World Congress*, 13–18 August 2017, Kuala Lumpur, pp. 4061–4069.

Pearson, T. H. and Rosenberg, R. (1978). 'Macrobenthic succession in relation to organic enrichment and pollution of the marine environment', in H. Barnes (ed.), *Oceanography and Marine Biology: An Annual Review* (vol 16). Aberdeen, UK: Aberdeen University Press, pp. 229–311.

Perelo, L. W. (2010). 'In situ and bioremediation of organic pollutants in aquatic sediments'. *Journal of Hazardous Materials*, vol 177, pp. 81–89.

Piña-Ochoa, E. and Álvarez-Cobelas, M. (2006). 'Denitrification in aquatic environments: A cross-system analysis'. *Biogeochemistry*, vol 81, pp. 111–130.

Pinckney, J. L., Paerl, H. W., Tester, P. and Richardson, T. L. (2001). 'The role of nutrient loading and eutrophication in estuarine ecology'. *Environmental Health Perspectives*, vol 109, pp. 699–706. doi:10.2307/3454916.

Quinn, J. M., Cooper, A. B., Davies-Colley, R. J., Rutherford, J. C. and Williamson, R. B. (1997). 'Land use effects on habitat, water quality, periphyton, and benthic invertebrates in Waikato, New Zealand, hill-country streams'. *New Zealand Journal of Marine and Freshwater Research*, vol 31, pp. 579–597.

Rippy, M. A., Stein, R., Sanders, B. F., Davis, K., McLaughlin, K., Skinner, J. F., Kappeler. J and Grant, S. B. (2014). 'Small drains, big problems: The impact of dry weather runoff on shoreline water quality at enclosed beaches'. *Environmental Science and Technology*, vol 48, no 24, pp. 14168–14177. doi:10.1021/es503139h.

Robertson, B. and Stevens, L. (2011). 'New river estuary macroalgal monitoring 2010/2011. Prepared for Environment Southland'. Report by Wriggle Coastal Management, Nelson, New Zealand.

Robertson, B. and Stevens, L. (2013). 'New river estuary fine scale monitoring of highly eutrophic arms 2012/2013. Prepared for Environment Southland'. Report by Wriggle Coastal Management, Nelson, New Zealand.

Roslev, P., Bastholm, S. and Iversen, N. (2008). 'Relationship between fecal indicators in sediment and recreational waters in a Danish estuary'. *Water, Air, and Soil Pollution*, vol 194, no 1–4, pp. 13–21.

Russell, V. S. (1974). 'Pollution: Concept and definition'. *Biological Conservation*, vol 6, no 3, pp. 157–161.

Schang, C., Lintern, A., Cook, P. L. M., Osborne, C., McKinley, A., Schmidt, J., Colman, R., Rooney, G., Henry, R., Deletica, A and McCarthy, D. (2016). 'Presence and survival of culturable *Campylobacter* spp. and *Escherichia coli* in a temperate urban estuary'. *Science of the Total Environment*, vol 569–570, pp. 1201–1211.

Seitzinger, S., Harrison, J. A., Bohlke, J. K., Bouwman, A. F., Lowrance, R., Peterson, B., Tobias, C. and Drecht, G. V. (2006). 'Denitrification across landscapes and waterscapes: A synthesis'. *Ecological Applications*, vol 16, pp. 2064–2090.

Tagg, A. S., Sapp, M., Harrison, J. P. and Ojeda, J. J. (2015). 'Identification and quantification of microplastics in wastewater using focal plane array-based reflectance micro-FT-IR imaging'. *Analytical Chemistry*, vol 87, no 12, pp. 6032–6040.

Talvitie, J., Heinonen, M., Paakkonen, J. P., Vahtera, E., Mikola, A., Setala, O. and Vahala, R. (2015). 'Do wastewater treatment plants act as a potential point source of microplastics? Preliminary study in the coastal Gulf of Finland, Baltic Sea'. *Water Science & Technology*, vol 72, no 9, pp. 1495–1504.

Thrush, S. F., Townsend, M., Hewitt, J. E., Davies, K., Lohrer, A. M., Lundquist, C. J. and Cartner, K. (2013). 'The many uses and values of estuarine ecosystems', in J. Dymond (ed.), *Ecosystem Services in New Zealand – Conditions and Trends*. Lincoln, New Zealand: Manaaki Whenua Press.

Tilman, D. (1999). 'Global environmental impacts of agricultural expansion: The need for sustainable and efficient practices'. *Proceedings of the National Academy of Sciences*, vol 96, pp. 5995–6000.

Villarrubia-Gómez, P., Cornell, S. E. and Fabres, J. (2018). 'Marine plastic pollution as a planetary boundary threat – The drifting piece in the sustainability puzzle'. *Marine Policy*, vol 96, pp.213–220. doi:10.1016/j.marpol.2017.11.035.

Vitousek, P. M., Aber, J. D., Howart, H. R. W., Likens, G. E., Matson, P. A., Schindler, D. W., Schlesinger, W. H. and Tilman, D. G. (1997). 'Human alteration of the global nitrogen cycle: Sources and consequences'. *Ecological Applications*, vol 7, pp. 737–750.

Ziajahromi, S., Neale, P. A. and Leusch, F. D. L. (2016). 'Wastewater treatment plant effluent as a source of microplastics: Review of the fate, chemical interactions and potential risks to aquatic organisms'. *Water Science & Technology*, vol 74, no 10, pp. 2253–2269.

Zillén, L., Conley, D. J., Andrén, T., Andrén, E. and Björck, S. (2008). 'Past occurrences of hypoxia in the Baltic Sea and the role of climate variability, environmental change and human impact'. *Earth Science Reviews*, vol 91, pp. 77–92.

8 Impacts of marine heatwaves

Belinda L Cannell, Paul G Thomson,
Verena Schoepf, Chari B Pattiaratchi and
Matthew W Fraser

Introduction

Marine research has often focussed on the ecological impacts of steadily changing conditions related to climate change such as gradual warming and gradual acidification. However, climate change will also lead to an increase in the frequency and intensity of extreme climatic events, primarily marine heatwaves, which have already led to detrimental impacts in marine ecosystems globally (Oliver *et al.* 2018). A marine heatwave is defined as a region of anomalously warm sea surface temperature (SST) that persists for at least five days (Hobday *et al.* 2016a). Such heatwaves have been observed globally including: north-east Pacific (California, Gulf of Alaska); north-west Atlantic (Gulf of Maine); coasts of Australia (Western Australia, Tasmania and the Great Barrier Reef); and the north-west Mediterranean. Furthermore, marine heatwaves are now being categorised to represent their variable severity (Hobday *et al.* 2018a). However, the processes that lead to these heatwaves are not well understood as they are due to a combination of different processes that occur simultaneously and are not common to each of the regions. In other words, different combinations of processes occur in different regions leading to a heatwave although many events coincide with large scale ocean oscillations such as ENSO (El Niño Southern Oscillation) and NAO (North Atlantic Oscillation) events that influence the air–sea heat exchange over an extended time period (weeks to months). For example, the extreme heatwave off Western Australia in 2010–2011 occurred under significant La Niña conditions that resulted in a strong, unseasonal Leeuwin Current in late summer/ autumn. Climate change scenarios (e.g. RCP8.5) indicate increasing ocean temperature and heat content as well as an increase in the strength of ocean boundary currents (Wu *et al.* 2012). In combination, these conditions are likely to lead to more frequent and intense marine heatwaves (Oliver *et al.* 2018). Furthermore, under the RCP8.5 scenarios for 2020–2040, marine heatwaves of >3°C due to anthropogenic climate change have a >90 per cent chance of occurring (Oliver *et al.* 2017).

So what ecological impacts do these temperature extremes have? All species function optimally within a specific temperature range, or thermal zone. Survival, reproduction, growth and ultimately population size and species distribution are therefore strongly associated with temperature (Gupta *et al.* 2015). The biological impacts of marine temperature extremes are varied; some are positive but many are negative. The impacts that have been observed around the world include the following:

1　outbreaks of disease, e.g. the viral-induced Pacific Oyster Mortality Syndrome (Oliver *et al.* 2017), and toxic blooms releasing unprecedented levels of a marine neurotoxin (domoic acid) that causes widespread contamination of fish and shellfish as well as the mortality of a range of marine mammals (McCabe *et al.* 2016);

2　increased mortality in a variety of species, e.g. kelp (Wernberg *et al.* 2012), abalone (Caputi *et al.* 2016), fish (e.g. Hobbs and McDonald 2010), sea lions, whales and seabirds (e.g. Frölicher and Laufkötter 2018; Jones *et al.* 2018);

3　coral bleaching (e.g. Depczynski *et al.* 2013);

4　changes in the phenology, or the timing, of life cycle events such as migration and breeding in relation to climatic variables (e.g. Mills *et al.* 2013);

5　changes in growth patterns such as faster growth rates of lobsters in warmer waters (Mills *et al.* 2013);

6　trophic mismatches, where the timing of the presence and abundance of prey no longer matches the needs of the predator (e.g. Edwards and Richardson 2004);

7　contraction of cooler waters, resulting in smaller spatial extent of prey and hence increased competition (Jones *et al.* 2018); and

8　range extensions and contractions of a diverse range of flora and fauna; such changes in range extensions can be short-lived (e.g. Smith *et al.* 2011) or permanent (e.g. Cheung *et al.* 2013), and can result in potential competition between newcomers and resident species and ultimately a shift in community structure (Wernberg *et al.* 2012), borne by the differential effects on species (Okey *et al.* 2014).

The species most vulnerable to extreme temperatures are those which have a narrow thermal zone. But which species are those? Because species can potentially acclimatise or adapt it is necessary to identify the heat tolerance of every species under different thermal regimes (Bennett *et al.* 2015) to determine the most vulnerable species. Yet, this seems to be an impossible task. Furthermore, there is also debate about whether species at the rear edge of their distribution are more likely to be impacted by a changing climate (e.g. Hampe and Petit 2005; Viejo *et al.* 2010), such as the projected rate of shift of their range (e.g. Poloczanska *et al.* 2013; Robinson *et al.* 2015). Additionally, some species at the centre of their range have been found to be as vulnerable as those at range edge (Bennett *et al.* 2015). It is also difficult to determine the most vulnerable ecosystems. This is because the impacts of extreme temperatures can occur over a range of temporal and spatial scales. Furthermore, the adaptive capacity and the sensitivity of the ecosystem underpin its vulnerability (Frölicher and Laufkötter 2018). While it is difficult to predict the species or ecosystems that are most vulnerable, the areas of fastest ocean warming have been identified, based on both historical data and modelling for future persistence. Globally, 24 marine 'hotspots' have been identified where warming is occurring 90 per cent faster than the rest of the oceans (Hobday and Pecl 2014). They cover a range of latitudes and longitudes, are mostly associated with western and eastern boundary currents, and some have occurred in areas where humans are very reliant on marine resources (Hobday and Pecl 2014). While these hotspots do not elucidate the most vulnerable ecosystems, they can be instrumental for the planning of effective management and adaptation efforts of the species found in those areas (Hobday and Pecl 2014).

Impacts on socio-ecological services

Understanding the impacts of marine heatwaves on species and ecosystems involves important scientific research. Beyond concerns for the species and ecological communities themselves, there are significant socio-cultural and socio-economic ramifications of marine temperature extremes given the heavy reliance humans place upon oceans and marine resources. Importantly, there are implications for food security as aquaculture and fisheries are key sectors that will be impacted (Hobday and Pecl 2014; Hobday *et al.* 2018b; Oliver *et al.* 2018). In many parts of the world up to 50 per cent of protein comes from fish and seafood (FAO 2016). Therefore, if marine food webs are disrupted this may result in a lack of food to satisfy nutritional requirements. In addition, there are other socio-cultural consequences including human security for communities that depend upon the ocean to supply resources (other than food) and artisanal livelihoods for local people. Furthermore, marine resources and the ocean environment itself provide the context for a range of cultural practices and recreational activities. Other socio-economic impacts include potential financial losses for industrial fisheries and coastal tourism which rely upon pristine marine environments, abundant fish stocks and healthy reefs.

These impacts can be felt at the local level but can also be at scales that affect national economies and therefore lead to international economic tensions, as well as providing motivation for maritime crimes as explored in this volume (see Chapter 11). From an environmental perspective, the loss of ecosystem services can impact whole countries and regions, physically and financially. For example, the loss of nutrient cycling by seagrass meadows has been estimated to cost US$29,000/ha/year. Similarly, the loss of coral reefs is estimated at US$352,000/ha/year (Costanza *et al.* 2014).

If coastal ecosystems are stressed by marine heatwaves this can negatively affect their ability to contribute to climate change adaptation initiatives.

Nature-based solutions to climate change rely upon ecosystem services provided, for example, by seagrass beds and coral reefs (see Chapter 5). National governments are required to commit to implementing climate change adaptation under the 2015 Paris Agreement and, if natural resources are to be relied upon, better understanding of the current and predicted effects of marine heatwaves is imperative.

As previously mentioned, a recent marine heatwave occurred along the coast of Western Australia. During the austral summer of 2010–2011, warmer than average sea surface temperatures were recorded over several weeks and covered approximately 600,000 km^2 (Pearce and Feng 2013). It had profound effects on a number of species and coastal habitats as well as people. This case study is used to explore the specific effects in further detail. We describe the physical drivers of this heatwave, and then exemplify its diverse areas of impact.

Case study: 2011 Western Australia marine heatwave

An unprecedented extreme marine heatwave event occurred along the 1,500 km coast off Western Australia in early 2011. This was coincident with a very strong La Niña event (Figure 8.1), and such events are associated with strong trade winds in the equatorial Pacific Ocean. This results in warmer surface water being pushed into the south-eastern Indian Ocean and a strengthening of the Leeuwin Current (Pearce and Feng 2013), which flows southwards along the west coast (Cresswell and Golding 1980) and is strongest in the winter (Feng *et al.* 2003). The unprecedented La Niña, in turn, increased the Leeuwin Current's volume transport in February – an unusual event at this time of the year that resulted in the southward transport of water that was ~5°C higher than normal (Feng *et al.* 2013). Other factors were also responsible for the development of the heatwave. These include a reduction in southerly winds along the west coast which resulted in a transfer of heat from the air into the ocean, and a multi-decadal increase in the strength of the Pacific trade winds (Feng *et al.* 2013). The elevated higher temperatures persisted in the coastal water for a period of 8–10 weeks (see Figure 8.1) and were identified as the most extreme warming event over a period of 140 years (Wernberg *et al.* 2012). Based on the example of the 2011 Western Australia marine heatwave, this chapter highlights strong and widespread impacts of temperature extremes on individual species and ecosystems.

Phytoplankton and zooplankton

The 2011 marine heatwave impacted the plankton through reduced primary productivity, changes in phytoplankton community composition and a decrease in zooplankton biomass. Thompson *et al.* (2015) compared biological and biogeochemical parameters along the West Australian coastline between the El Niño year of 2010 and the La Niña heatwave in 2011. Using monitoring data from Australia's Integrated Marine Observing System

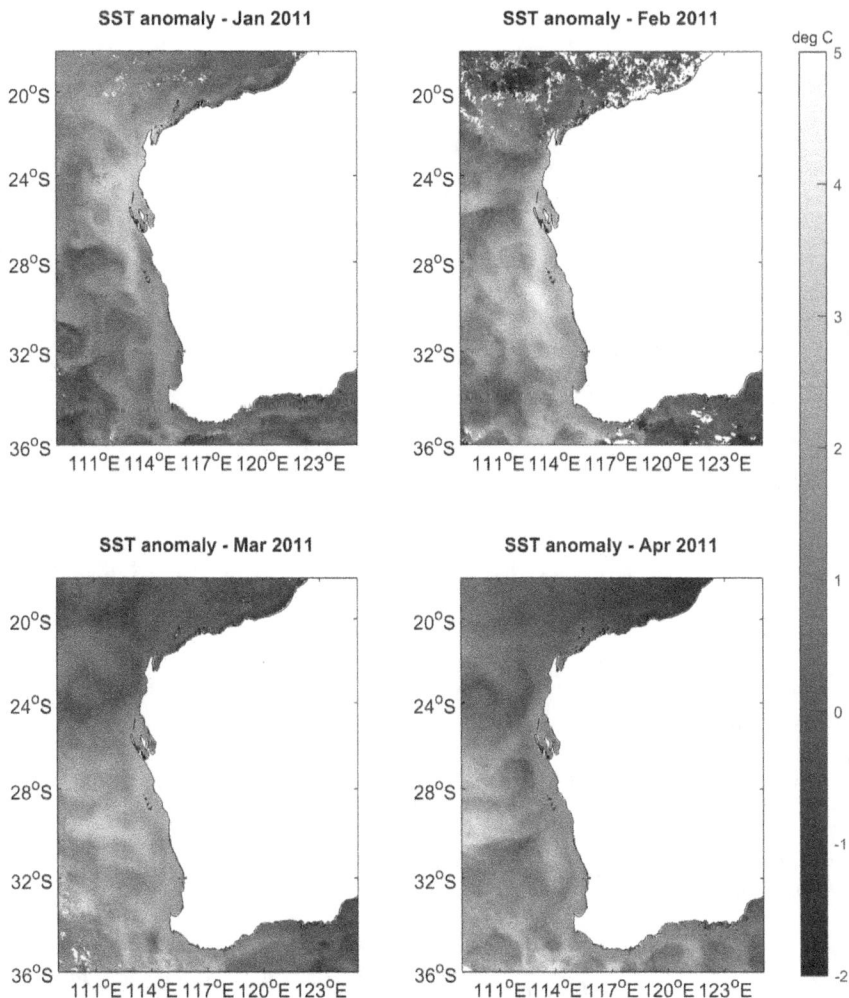

Figure 8.1 The size and movement of the marine heatwave off Western Australia, January–April 2011.

Source: Chari Pattiaratchi.

Notes
SST – sea surface temperature; the grey shades indicate water temperature deviations above normal.

(IMOS) at National Reference Stations (NRS; see Chapter 6) at Ningaloo Reef, Rottnest Island and Esperance (Figure 8.2), they found a localised reduction in chlorophyll *a* along some areas of the north-west coast and an overall reduction in primary productivity. In the south-west of Western Australia, at Rottnest Island near Perth, higher nutrient concentrations (nitrate

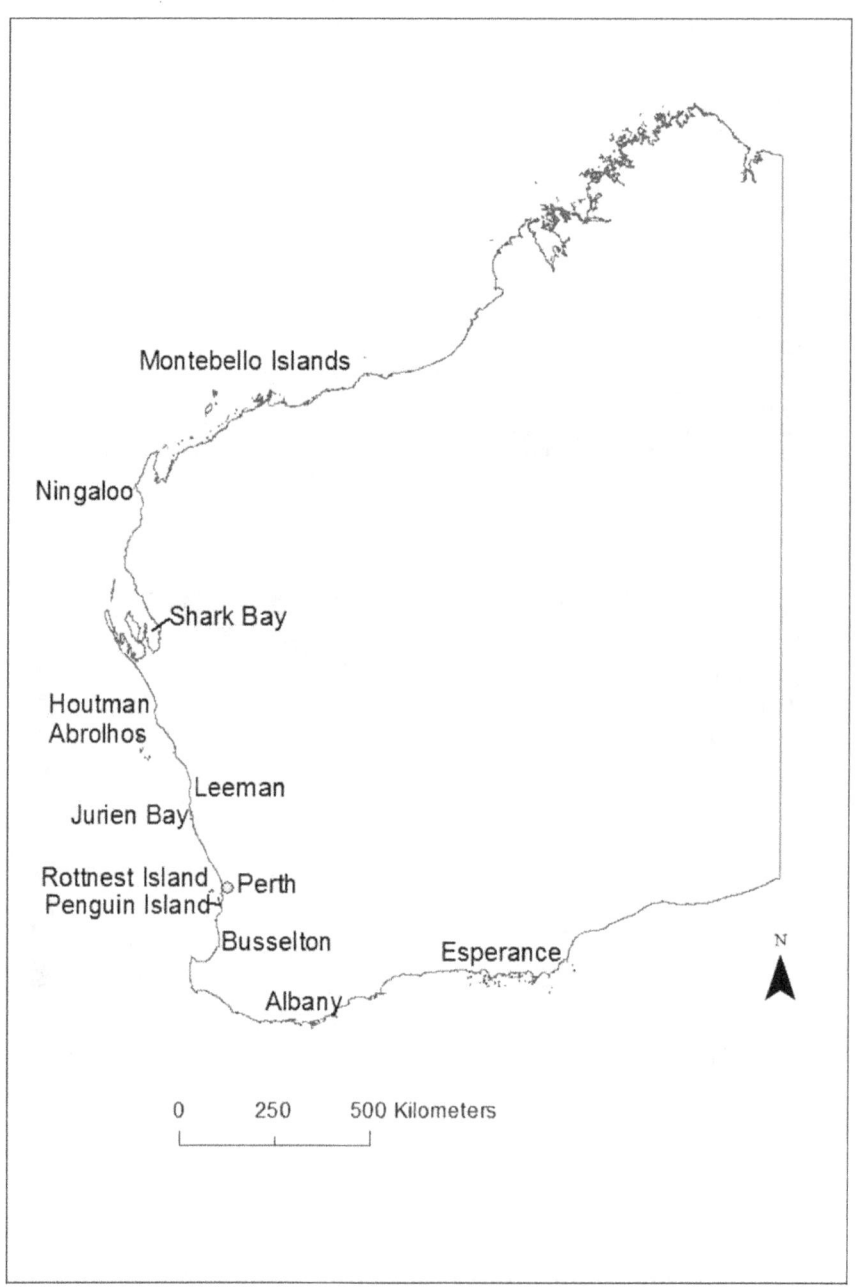

Montebello Islands

Ningaloo

Shark Bay

Houtman
Abrolhos

Leeman
Jurien Bay

Rottnest Island ₒPerth
Penguin Island

Busselton   Esperance   N

Albany

0   250   500 Kilometers

Figure 8.2 Map of Western Australia, indicating the areas of impact exemplified in this chapter.

Source: Belinda L. Cannell.

and silicate) were available for phytoplankton growth in the upper sunlit layers during the heatwave. This increase in nutrients may have resulted in a change in phytoplankton community composition with an increased abundance of the very small and tropical picoplankton (*Prochlorococcus* and *Synechococcus*) recorded during the heatwave. Along the south coast at Esperance, phytoplankton community composition was also altered during the marine heatwave, where an increased proportion of the photosynthetic pigment 19-hexanoyloxyfucoxanthin indicated a greater abundance of small coccolithophorids in this region. At each IMOS NRS station along the Western Australian coastline, Thompson *et al.* (2015) also found that zooplankton biomass was lower during the heatwave compared to the previous year, although not significantly.

What are the possible consequences of these changes in the plankton? Unlike the dramatic and highly visible effects of the 2011 marine heatwave on the larger and more visible species, as outlined below, the effects on plankton may be far more insidious. The plankton form the base of the marine food chain and a shift in community composition towards the smaller, more difficult to harvest and less nutritious cells such as the picoplankton and the coccolithophorids may have a cascade effect on higher trophic levels. For example, it is possible that changes in the plankton indirectly impacted breeding and breeding success of Little Penguins (discussed below) though deleterious effects on the prey species of the penguins. While such effects are difficult to measure, it is likely that the changes observed in the plankton communities place additional stresses on already temperature stressed higher trophic levels during marine heatwaves.

Corals

Coral reefs throughout the world have been impacted by increasingly frequent and severe mass bleaching events since the 1980s, most notably during strong El Niño events such as 1982–1983, 1987, 1997–1998, 2010 and most recently 2015–2016 (Hughes *et al.* 2018a). Coral bleaching occurs when stressful conditions, such as unusually warm temperatures, disrupt the mutually beneficial relationship (i.e. symbiosis) between corals and their dinoflagellate symbionts (*Symbiodinium* spp.). Since corals depend on their symbionts to meet their energy demand, bleaching leads to severe resource limitation, increased susceptibility to disease and often high levels of mortality (Hoegh-Guldberg 1999; Hughes *et al.* 2018a). However, with the exception of some offshore reefs (Gilmour *et al.* 2013), coral reefs in Western Australia largely escaped mass bleaching until the 2011 marine heatwave, which caused unprecedented coral bleaching and mass mortality across 12 degrees of latitude (Moore *et al.* 2012). Since 2011, more localised bleaching events have continued to impact coral reefs in Western Australia, but mass bleaching did not occur until the strong El Niño of 2015–2016 which caused regional-scale mass bleaching in the north-west (Le Nohaïc *et al.* 2017).

Mass bleaching during the 2011 marine heatwave extended from the Montebello Islands and Ningaloo Reef all the way down to subtropical coral communities in the Houtman Abrolhos Islands and Rottnest Island off the coast of Perth (see Fig 8.2) (Moore *et al.* 2012). At the Montebello Islands, between 34 and 77 per cent of corals bleached but coral mortality was relatively moderate and localised (Moore *et al.* 2012). In contrast, the Ningaloo Reef region was severely impacted by bleaching (up to 95 per cent in the Gulf of Exmouth) (Moore *et al.* 2012), and also experienced extensive coral mortality across the region, resulting in the greatest proportional loss of coral cover (32–92 per cent) of all regions impacted by the heatwave (Moore *et al.* 2012; Depczynski *et al.* 2013). Bleaching at the Houtman Abrolhos and at Rottnest Island was comparatively moderate with 12–22 per cent and 17 per cent, respectively, but resulted in relatively high (23–50 per cent) loss of coral cover at the Houtman Abrolhos (Abdo *et al.* 2012; Moore *et al.* 2012; Smale and Wernberg 2012). At these high-latitude sites, severe and extensive bleaching extended down to deeper (>20 m) coral communities (Thomson *et al.* 2011; Smale and Wernberg 2012). This highlighted that even high-latitude reef locations may not be considered refugia from climate change.

Comparison of the geographic footprint of the 2011 and 2016 mass bleaching events shows that coral reefs in central-to-southern Western Australia are primarily at risk of severe bleaching during strong La Niña events, such as in 2011, whereas reefs in northern Western Australia are at risk during strong El Niño events, such as in 2016 (Le Nohaïc *et al.* 2017; Zhang *et al.* 2017). However, as ocean warming intensifies, mass bleaching events are increasingly occurring during all El Niño-Southern Oscillation phases (Hughes *et al.* 2018b). Given that millions of people rely on coral reefs for their livelihoods and ecological services (Moberg and Folke 1999; Spalding *et al.* 2017), marine heatwaves and associated mass bleaching events have immense socio-economic implications, as noted above, particularly for coastal communities that are dependent upon healthy reefs for their protein intake, protection of coastlines from storms and to support tourism enterprises.

Seagrasses

Seagrasses (marine angiosperms) are another group of marine primary producers that were negatively impacted by the 2011 marine heatwave. This was particularly true in Shark Bay (see Figure 8.2), which contains one of the largest seagrass meadows in the world (4,300 km^2). The seagrass species *Amphibolis antarctica* was particularly impacted in the Shark Bay World Heritage Area, with mass defoliation and dieback recorded in the months following the heatwave. Though leaf biomass increased in the two years following the heatwave, below ground biomass (roots and rhizomes) decreased, eroding the resilience of the surviving seagrasses to future warming events (Fraser *et al.* 2014). Up to 1,100 km^2 (~36 per cent) of seagrass was damaged following the heatwave (Arias-Ortiz *et al.* 2018).

The loss of seagrasses in Shark Bay had bottom-up impacts on the remainder of the ecosystem. Meadows previously occupied by the habitat-forming, large bodied *A. antarctica* have been replaced by the smaller, tropical seagrass *Halodule uninervis* (Nowicki *et al.* 2017), reducing the foraging habitat for megafauna such as green turtles. Indeed, in the years following the heatwave, a significant reduction in green turtle health was recorded in Shark Bay (Thomson *et al.* 2015). The seagrass meadows of Shark Bay also sequester a large volume of organic carbon in sediments (Fourqurean *et al.* 2012). However, the proportion of organic carbon stored in sediments under Shark Bay seagrasses has decreased dramatically since the 2011 marine heatwave, with between 2 and 9 Tg CO_2 released into the atmosphere following the loss of seagrass; this has increased CO_2 emissions from land-use change in Australia by between 4 and 21 per cent per annum (Arias-Ortiz *et al.* 2018).

Significant attention is now given to seagrasses as they provide ecosystem services by acting as carbon sinks. Initiatives based upon blue carbon offsets are being promoted as climate change mitigation strategies as part of commitments under the 2015 Paris Agreement (e.g. The Blue Carbon Initiative, http://thebluecarboninitiative.org/blue-carbon/). Negative impacts upon seagrasses may therefore inhibit the achievement of national and international greenhouse gas mitigation goals.

Macroalgae

Several species of macroalgae were also impacted by the 2011 marine heatwave (Wernberg *et al.* 2016). The brown algae *Scytothalia dorycarpa* underwent a range contraction of around 100 km after the heatwave, representing a loss of about 5 per cent of its global distribution (Smale and Wernberg 2013). This mirrored broader changes in understorey taxa, with increases in turf-forming algae and red macroalgae and decreases in encrusting coralline algae, encrusting non-coralline algae, sponges and articulate coralline algae (Smale and Wernberg 2013). The kelp species *Ecklonia radiata* was also impacted across its Western Australian range, with unprecedented fouling of kelp at the Houtman Abrolhos (Smale and Wernberg 2012) and reduced cover at Jurien Bay (see Figure 8.2), where it was largely replaced by turf-forming algae. The total cover of macroalgal canopy (kelps and large fucoids) decreased after the heatwave, highlighting the impacts to several habitat forming marine species.

As with seagrass and corals, the loss of macroalgae had bottom-up impacts in impacted ecosystems. Areas previously dominated by temperate seaweeds and fishes were broadly replaced by warm-water species (Wernberg *et al.* 2016). The subtropical and tropical fish species that replaced the temperate fish contributed to over-grazing on canopy seaweeds, preventing recovery and leading to a regime shift from marine ecosystems dominated by kelp forests to those dominated by turf algae. The long-term prospects of kelp forests in Western Australia are under threat, with more frequent warming events predicted to further decrease the geographical range of these critical marine ecosystems, threatening associated fishing and tourism industries that are estimated to be worth more than AUD$10 billion per year (Bennett *et al.* 2016).

Fish

Another phenomenon associated with the 2011 heatwave was the tropicalisation (i.e. the appearance of warmer-water species in typical temperate zones) of fish communities in coastal habitats. While sessile organisms have limited opportunity to move to refugia to escape mortality during heatwaves, most fish species have the ability to locate cooler conditions within their physiological tolerances. Other tropical species that can physiologically tolerate warmer temperatures arrived in southern regions. For example, in Jurien Bay there was an increase in warm-water species such as Western Scalyfin (*Parma occidentalis*), Western Butterflyfish (*Chaetodon assarius*) and Lined Dottyback (*Labracinus lineatus*). At a community level, the ratio of tropical to non-tropical species doubled after the event (from 5–10 per cent to 20 per cent) (Wernberg *et al.* 2012), showing the capacity of heatwaves to drive changes in fish community composition. Further south, there were record damselfish numbers at Rottnest Island following the heatwave. Six of these damselfish species represented the first time the species had been found at Rottnest (Pearce *et al.* 2011). At Busselton (see Figure 8.2), several new species were spotted in the area, including raccoon butterflyfish (*Chaetodon lunula*), reef bannerfish (*Heniochus diphreutes*) and Scissortail Sergeants (*Abudefduf sexfasciatus*). However, there were no sightings of short-tailed nudibranchs (*Ceratosoma brevicaudatum*), down from 74–80 per cent sightings between March 2006 and March 2008 (Pearce *et al.* 2011). Manta Rays (*Manta birostris*, normally distributed in the tropics) were reported as far south as Albany (Pearce *et al.* 2011; Lenanton *et al.* 2017). The variability in species impacted, and subsequent shifts in community composition, underlines the difficulties in predicting and managing the impacts of marine heatwaves.

As well as community level changes, there were also behavioural changes in several fish and elasmobranch species during the heatwave. For example, Western Australian Salmon (*Arripis truttaceus*) had increased feeding rates and changed feeding preferences (Pearce *et al.* 2011), Western Rock Lobsters (*Panulirus cygnus*) were noted to act erratically by moving away from the protection of the reef during the day at the Houtman Abrolhos (Pearce *et al.* 2011), while Wobbegongs (*Orectolobus ornatus*) were noted venting gills at the bottom of the sea floor (Pearce *et al.* 2011). However, the full ecological impacts of such behaviours are rarely studied. Mass fish kills were recorded at the Houtman Abrolhos, and from Moore River to Green Head, and were linked to rapid changes in sea temperature (Pearce *et al.* 2011). Many fish kills were reported after fish washed up on beaches or were found floating in coastal areas. Fish kills were dominated by benthic species (e.g. Wrasse, Cobbler, Leatherjacket) but some mobile species were also impacted (whiting, herring, mullet). In the Houtman Abrolhos, the species impacted by the fish kills were varied, with both large and small species impacted, including iconic species such as Baldchin Grouper (*Choerodon rubescens*) and Breaksea Cods (*Epinephelides armatus*).

Commercially important fishery species were also widely impacted by the heatwave. Western Rock lobster – the single most valuable fishery in Western Australia – experienced widespread mortality at major fishery grounds such as the Houtman Abrolhos and Leeman (Pearce *et al.* 2011). Roe's abalone (*Halitosis roei*) were decimated north of the Murchison River, with no commercial fishing survey finding a live individual in the area between March and June 2011. This led managers to close an area of the commercial fishery and the recreational fishery to encourage stock recovery (Pearce *et al.* 2011). The scallop fishery in Shark Bay was also impacted with reduced growth of scallops (*Amusium balloti*) and delayed timing of larval settlement. Other fisheries were positively impacted – there was an increased survival of Western King Prawns (*Penaeus latisulcatus*) and Brown Tiger Prawns (*Penaeus esculentus*) in Shark Bay and Exmouth Gulf, and Blue Swimmer Crabs (*Portunus armatus*) in Exmouth Gulf were more abundant. This had positive impacts on fisheries. For example, the number of Brown Tiger Prawns landed in the Exmouth Gulf fishery was one of the highest on record in 2011 following the heatwave, though this was followed by lower recruitment in subsequent years, thought to be driven by loss of seagrass habitats (Caputi *et al.* 2015). Recreational fisheries were negatively impacted, with a major failure in the seasonal Australian Salmon fishery which is a major tourism attraction (Pearce *et al.* 2011). Again, the variability of responses between fisheries stocks underlines the challenges that marine heatwaves will increasingly pose for the prediction and management of stocks as we move into a period where such events become more frequent and intense (Oliver *et al.* 2018).

Penguins

The northern-most colony of Little Penguins (*Eudyptula minor*) in Western Australia is located on Penguin Island, 50 km south of Perth (see Figure 8.2). Long term monitoring in the region (e.g. Klomp and Wooller 1988; Murray *et al.* 2011; Cannell *et al.* 2012; Cannell *et al.* 2013; Cannell *et al.* 2016) has been critical to evaluating the effects of the marine heatwave.

The impact of the 2011 marine heatwave on the penguins was marked, with a reduction in both the number of penguins participating in breeding and the success of the breeding attempts (Cannell *et al.* 2012). There was a contraction in the breeding season in 2011 (Cannell *et al.* 2012), giving less opportunity for the penguins to raise two clutches in the year. The penguins were lighter than average (1380 g vs 1290 g for males and 1230 g vs 1160 g for females prior to 2011, and during 2011, respectively), and the body condition of the penguins impacts their ability to breed as well as successfully undergo their annual moult. Their usual major dietary component during the breeding season, Whitebait (*Hyperlophus vittatus*), was replaced with Scaly Mackerel (*Sardinella lemuru*) (Cannell 2012), a tropical fish species that had previously rarely been found in their diet. There was also a four-fold increase in the number of dead penguins, with many dying from starvation (Cannell *et al.* 2016). The impact of the heatwave continues, however, due partly to the fact that penguins are faithful to the natal colony, with both adults and newly mature 2-year-old and 3-year-old individuals returning each year. But with fewer adults breeding during the marine heatwave, and fewer chicks surviving, there are fewer juveniles returning to boost the population. Increasing frequency and intensity of heatwaves is likely to have a negative long-term impact on the penguins on Penguin Island.

Pest species

Invasions by marine species can have both ecological and economic consequences. The growth and success of invasive species is correlated with increasing water temperature (McDonald 2012). Thus, marine heatwaves have the potential to provide a window of suitable conditions for pests that are not considered a risk in a particular region. One of the most commonly found invasive species on vessels entering Western Australian waters is the Asian Green mussel (*Perna viridis*). It is not yet established in Australia, but it is feared that it could become as problematic as the invasive freshwater mussel *Dreissena polymorpha* in Europe and North America, which has cost nearly US$1 billion to manage (Power *et al.* 2004). The Asian Green mussel can tolerate a wide range of salinities, water temperatures and turbidity (Power *et al.* 2004). It not only rapidly outcompetes native mussels but also fouls hard surfaces, with adult densities up to $4,000/m^2$ (Power *et al.* 2004). Following the marine heatwave, an individual mussel was discovered on the hull of one vessel and a founder population which included many juveniles was discovered in the sea chest of another vessel. Both vessels were berthed at the naval base, Garden Island (McDonald 2012). Spawning was thought to be triggered by the >3°C SST experienced during the marine heatwave (McDonald 2012). The mussels were removed and follow-up surveys of nearby hard structures found no evidence of them (McDonald 2012). This highlights the importance of vigilance regarding monitoring for invasive species, which have the capacity to wreak havoc on a range of ecosystems and the services they provide.

Conclusion

Temperature extremes in the form of marine heatwaves have marked impacts on marine ecosystems globally, with range extensions and contractions of species distribution, and ecosystem-wide changes. This will naturally impact the ecosystem services that the ocean provides, as has already been demonstrated with the case study presented here. Given these services are fundamental to human well-being, it behoves action at multiple levels, from individual citizens to global policy makers. While mitigation is the long-term goal, adaptation is needed at local scales.

Scientists have the most critical role to play in better understanding the effects of marine heatwaves on species and ecosystems, and thereafter to predict future impacts and ramifications. This critical role of scientists has been demonstrated for the fisheries and aquaculture sector, where forecasting the impact of seasonal and long term climate change can now be used to adjust management practices (Hobday *et al.* 2018b). It is also critical that this knowledge is communicated in accessible ways to governments, industry and communities. The categorisation of marine heatwaves is helpful in this regard. Governments must also ensure that the scientific knowledge is embedded in law, policy and decision-making processes, to enhance conservation and utilisation in the light of climate change and marine heatwaves of varied intensity and frequency. Such efforts on a governmental level must ensure human and food security where impacts are likely and/or significant. Similarly, industry can use this information in strategic planning for sustainable future operations, as exemplified by the use of seasonal forecasting of rainfall and water temperatures for stocking and harvesting prawn aquaculture ponds (Hobday *et al.* 2016b). Communities, particularly in coastal areas, have an inherent interest in understanding impacts on their local marine environment, in order to safeguard their livelihoods and to support government responses to climate change (see Chapter 4). Long-term monitoring of the ocean is critical as it ensures that the effects of marine heatwaves are put in context.

While mitigating the extent and severity of marine heatwaves is a wider societal issue and a long-term undertaking, we recommend the following to guide governments, industry and communities in future planning:

1 the adoption of a marine heatwave categorisation system to compare and contrast the extent and severity of these events worldwide, now and into the future (Hobday *et al.* 2018a);
2 the adoption of a simple naming convention for heatwaves to enhance scientific understanding and public awareness of these events (Hobday *et al.* 2018a);
3 the development and use of seasonal forecast and long-term climate models coupled with a decision-making framework to aid in future decision making, particularly by marine industries (e.g. Hobday *et al.* 2018b);
4 the establishment and/or continued funding and support for ocean observing systems that provide long term monitoring data (e.g. IMOS in Australia);
5 the development and/or support of collaborative programmes to investigate the current health of the marine hotspots where warming is occurring most rapidly, with engagement from communities, industry, government and research institutions.

This chapter has explored a case study in Western Australia to highlight critical impacts and issues which are shared among many marine environments globally. Only with such enhanced understanding of the effects of marine heatwaves can all stakeholders take appropriate action to mitigate the impacts upon their environment and operations.

References

Abdo, DA, Bellchambers, LM and Evans, SN (2012). 'Turning up the heat: Increasing temperature and coral bleaching at the high latitude coral reefs of the Houtman Abrolhos Islands'. *PLoS ONE*, vol 7, no 8, p. e43878, doi:10.1371/journal. pone.0043878.

Arias-Ortiz, A, Serrano, O, Masqué, P, Lavery, PS, Mueller, U, Kendrick, GA, Rozaimi, M, Esteban, A, Fourqurean, JW, Marbà, N and Mateo, MA (2018). 'A marine heatwave drives massive losses from the world's largest seagrass carbon stocks'. *Nature Climate Change*, vol 8, no 4, pp. 338–344. doi:10.1038/s41558-018-0096-y.

Bennett, S, Wernberg, T, Joy, BA, De Bettignies, T and Campbell, AH (2015). 'Central and rear-edge populations can be equally vulnerable to warming'. *Nature Communications*, vol 6, p. 10280. doi:10.1038/ncomms10280.

Bennett, S, Wernberg, T, Connell, SD, Hobday, AJ, Johnson, CR and Poloczanska, ES (2016). 'The "Great Southern Reef": Social, ecological and economic value of Australia's neglected kelp forests'. *Marine and Freshwater Research*, vol 67, pp. 47–56.

Cannell, BL (2012). 'Study of the effect of a boat ramp at Becher Point on the food resources of Little Penguins on Penguin Island: Little Penguin component'. Report for The Department of Environment and Conservation, Canberra, Australia.

Cannell, BL, Chambers, LE, Wooller, RD and Bradley, JS (2012). 'Poorer breeding by Little Penguins near Perth, Western Australia is correlated with above average sea surface temperatures and a stronger Leeuwin Current'. *Marine & Freshwater Research*, vol 63, pp. 914–925.

Cannell, BL, Murray, D and Bunce, M (2013). 'The diet composition of Little Penguins from Penguin Island 2012'. Report for Department of Fisheries, Perth, Western Australia.

Cannell, BL, Campbell, K, Fitzgerald, L, Lewis, JA, Baran, IJ and Stephens, NS (2016). 'Anthropogenic trauma is the most prevalent cause of mortality in Little Penguins (*Eudyptula minor*) in Perth, Western Australia'. *Emu*, vol 116, pp. 52–61.

Caputi, N, Feng, M, Pearce, A, Benthuysen, J, Denham, A, Hetzel, Y, Matear, R, Jackson, G, Molony, B, Joll, L and Chandrapavan, A (2015). 'Management implications of climate change effect on fisheries in Western Australia Part 1: Environmental change and risk assessment'. FRDC Project No. 2010/535, Fisheries Research Report No. 260, Department of Fisheries, Perth, Western Australia.

Caputi, N, Kangas, M, Denham, A, Feng, M, Pearce, A, Hetzel, Y and Chandrapavan, A (2016). 'Management adaptation of invertebrate fisheries to an extreme marine heat wave event at a global warming hot spot'. *Ecology and Evolution*, vol 6, no 11, pp. 3583–3593.

Cheung, WW, Watson, R and Pauly, D (2013). 'Signature of ocean warming in global fisheries catch'. *Nature*, vol 497, no 7449, pp. 365–368.

Costanza, R, de Groot, R, Sutton, P, Van der Ploeg, S, Anderson, SJ, Kubiszewski, I, Farber, S and Turner, RK (2014). 'Changes in the global value of ecosystem services'. *Global Environmental Change*, vol 26, pp. 152–158.

Cresswell, GR and Golding, TJ (1980). 'Observations of a south-flowing current in the southeastern Indian Ocean'. *Deep Sea Research Part A. Oceanographic Research Papers*, vol 27, no 6, pp. 449–466.

Depczynski, M, Gilmour, JP, Ridgway, T, Barnes, H, Heyward, AJ, Holmes, TH, Moore, JAY, Radford, BT, Thomson, DP, Tinkler, P and Wilson, SK (2013). 'Bleaching, coral mortality and subsequent survivorship on a West Australian fringing reef'. *Coral Reefs*, vol 32, pp. 233–238.

Edwards, M and Richardson, AJ (2004). 'Impact of climate change on marine pelagic phenology and trophic mismatch'. *Nature*, vol 430, no 7002, pp. 881–883.

FAO (2016). 'The state of world fisheries and aquaculture 2016. Contributing to food security and nutrition for all'. Food and Agriculture Organization of the United Nations, Rome.

Feng, M, Meyers, G, Pearce, A and Wijffels, S (2003). 'Annual and interannual variations of the Leeuwin Current at 32°S'. *Journal of Geophysical Research: Oceans*, vol 108, no C11, p. 3355. doi:10.1029/2002JC001763.

Feng, M, McPhaden, MJ, Xie, SP and Hafner, J (2013). 'La Niña forces unprecedented Leeuwin Current warming in 2011'. *Scientific Reports*, vol 3, p. 1277. doi:10.1038/srep01277.

Fourqurean, JW, Kendrick, GA, Collins, LS, Chambers, RM and Vanderklift, MA (2012). 'Carbon, nitrogen and phosphorus storage in subtropical seagrass meadows: Examples from Florida Bay and Shark Bay'. *Marine and Freshwater Research*, vol 63, no 11, pp. 967–983.

Fraser, MW, Kendrick, GA, Statton, J, Hovey, RK, Zavala-Perez, A, Walker, DI and Lee, J (2014). 'Extreme climate events lower resilience of foundation seagrass at edge of biogeographical range'. *Journal of Ecology*, vol 102, pp. 1528–1536.

Frölicher, TL and Laufkötter, C (2018). 'Emerging risks from marine heat waves'. *Nature Communications*, vol 9, no 1, p. 650. doi:10.1038/s41467-018-03163-6.

Gilmour, JP, Smith, LD, Heyward, AJ, Baird, AH and Pratchett, MS (2013). 'Recovery of an isolated coral reef system following severe disturbance'. *Science*, vol 340, pp. 69–71.

Gupta, AS, Brown, JN, Jourdain, NC, van Sebille, E, Ganachaud, A and Vergés, A (2015). 'Episodic and non-uniform shifts of thermal habitats in a warming ocean'. *Deep Sea Research Part II: Topical Studies in Oceanography*, vol 113, pp. 59–72.

Hampe, A and Petit, RJ (2005). 'Conserving biodiversity under climate change: The rear edge matters'. *Ecology Letters*, vol 8, no 5, pp. 461–467.

Hobbs, JP and McDonald, CA (2010). 'Increased seawater temperature and decreased dissolved oxygen triggers fish kill at the Cocos (Keeling) Islands, Indian Ocean'. *Journal of Fish Biology*, vol 77, no 6, pp. 1219–1229.

Hobday, AJ and Peel, GT (2014). 'Identification of global marine hotspots: Sentinels for change and vanguards for adaptation action'. *Reviews in Fish Biology and Fisheries*, vol 24, no 2, pp. 415–425.

Hobday, AJ, Alexander, LV, Perkins, SE, Smale, DA, Straub, SC, Oliver, EC, Benthuysen, JA, Burrows, MT, Donat, MG, Feng, M and Holbrook, NJ. (2016a). 'A hierarchical approach to defining marine heatwaves'. *Progress in Oceanography*, vol 141, pp. 227–238.

Hobday, AJ, Spillman, CM, Eveson, JP and Hartog, JR (2016b). 'Seasonal forecasting for decision support in marine fisheries and aquaculture'. *Fisheries Oceanography*, vol 25, no S1, pp. 45–56.

Hobday, AJ, Oliver, ECJ, Gupta, AS, Benthuysen, JA, Burrows, MT, Donat, MG, Holbrook, NJ, Moore, PJ, Thomsen, MS, Wernberg, T and Smale, DA (2018a). 'Categorizing and naming marine heatwaves'. *Oceanography*, vol 31, no 2, pp. 1–13. doi:10.5670/oceanog.2018.5205.

Hobday AJ, Spillman CM, Eveson JP, Hartog JR, Zhang X and Brodie S (2018b). 'A framework for combining seasonal forecasts and climate projections to aid risk management for fisheries and aquaculture'. *Frontiers in Marine Science*, vol 5, no 137. doi:10.3389/fmars.2018.00137.

Hoegh-Guldberg, O. (1999). 'Climate change, coral bleaching and the future of the world's coral reefs'. *Marine and Freshwater Research*, vol 50, no 8, pp. 839–866.

Hughes, TP, Kerry, JT, Baird, AH, Connolly, SR, Dietzel, A, Eakin, CM, Heron, SF, Hoey, AS, Hoogenboom, MO, Liu, G, McWilliam, MJ, Pears, RJ, Pratchett, MS, Skirving, WJ, Stella, JS and Torda, G (2018a). 'Global warming transforms coral reef assemblages'. *Nature*, vol 556, no 7702, pp. 492–496.

Hughes, TP, Anderson, KD, Connolly, SR, Heron, SF, Kerry, JT, Lough, JM, Baird, AH, Baum, JK, Berumen, ML, Bridge, T, Claar, DC, Eakin, CAM, Gilmour, JP, Graham, NAJ, Harrison, H, Hobbs, JPA, Hoey, AS, Hoogenboom, MO, Lowe, RJ, McCulloch, M, Pandolfi, JM, Pratchett, MS, Schoepf, V, Torda, G and Wilson, SK (2018b). 'Spatial and temporal patterns of mass bleaching of corals in the Anthropocene'. *Science*, vol 359, pp. 80–83.

Jones, T, Parrish, JK, Peterson, WT, Bjorkstedt, EP, Bond, NA, Ballance, LT, Bowes, V, Hipfner, JM, Burgess, HK, Dolliver, JE and Lindquist, K (2018). 'Massive mortality of a planktivorous seabird in response to a marine heatwave'. *Geophysical Research Letters*, vol 45, no 7, pp. 3193–3202. doi:10.1002/2017GL076164.

Klomp, NI and Wooller, RD (1988). 'Diet of little penguins *Eudyptula minor* from Penguin Island, Western Australia'. *Australian Journal of Marine and Freshwater Research*, vol 39, pp. 633–640.

Le Nohaïc, M, Ross, CL, Cornwall, CE, Comeau, S, Lowe, R, McCulloch, MT and Schoepf, V (2017). 'Marine heatwave causes unprecedented regional mass bleaching of thermally resistant corals in northwestern Australia'. *Scientific Reports*, vol 7, no 1. doi:10.1038/s41598-017-14794-y.

Lenanton, RCJ, Dowling, CE, Smith, KA, Fairclough, DV and Jackson, G. (2017). 'Potential influence of a marine heatwave on range extensions of tropical fishes in the eastern Indian Ocean – Invaluable contributions from amateur observers'. *Regional Studies in Marine Science*, vol 13, pp. 19–31.

McCabe, RM, Hickey, BM, Kudela, RM, Lefebvre, KA, Adams, NG, Bill, BD, Gulland, FMD, Thomson, RE, Cochlan, WP and Trainer, VL (2016). 'An unprecedented coastwide toxic algal bloom linked to anomalous ocean conditions'. *Geophysical Research Letters*, vol 43, no 19, pp. 10366–10376.

McDonald, JI (2012). 'Detection of the tropical mussel species *Perna viridis* in temperate Western Australia: Possible association between spawning and a marine heat pulse'. *Aquatic Invasions*, vol 7, no 4, pp. 483–490.

Mills, KE, Pershing, AJ, Brown, CJ, Chen, Y, Chiang, F-S, Holland, DS, Lehuta, S, Nye, JA, Sun, JC, Thomas, AC and Wahle, RA. (2013). 'Fisheries management in a changing climate: Lessons from the 2012 ocean heat wave in the Northwest Atlantic'. *Oceanography*, vol 26, no 2, pp. 191–195. doi:10.5670/oceanog.2013.27.

Moberg, F and Folke, C (1999). 'Ecological goods and services of coral reef ecosystems'. *Ecological Economics*, vol 29, pp. 215–233.

Moore, JAY, Bellchambers, LM, Depczynski, MR, Evans, RD, Evans, SN, Field, SN, Friedman, KJ, Gilmour, JP, Holmes, TH, Middlebrook, R, Radford, BT, Ridgway, T, Shedrawi, G, Taylor, H, Thomson, DP and Wilson, SK (2012). 'Unprecedented mass bleaching and loss of coral across 12° of latitude in Western Australia in 2010–11'. *PLoS ONE*, vol 7, p. e51807. doi:51810.51371/journal.pone.0051807.

Murray, DC, Bunce, M, Cannell, BL, Oliver, R, Houston, J, White, NE, Barrero, RA, Bellgard, MI and Haile, J (2011). 'DNA-based faecal dietary analysis: A comparison of qPCR and throughput sequencing approaches'. *PLoS ONE*, vol 61, p. e25776. doi:25710.21371/journal.pone.0025776.

Nowicki, RJ, Thomson, JA, Burkholder, DA, Fourqurean, JW and Heithaus, MR (2017). 'Predicting seagrass recovery times and their implications following an extreme climate event'. *Marine Ecology Progress Series*, vol 567, pp. 79–93.

Okey, TA, Alidina, HM, Lo, V and Jessen, S (2014). 'Effects of climate change on Canada's Pacific marine ecosystems: A summary of scientific knowledge'. *Reviews in Fish Biology and Fisheries*, vol 24, no 2, pp. 519–559.

Oliver, ECJ, Benthuysen, JA, Bindoff, NL, Hobday, AJ, Holbrook, NJ, Mundy, CN and Perkins-Kirkpatrick, SE (2017). 'The unprecedented 2015/16 Tasman Sea marine heatwave'. *Nature Communications*, vol 8, p. 16101. doi:16110.11038/ncomms16101.

Oliver, EC, Donat, MG, Burrows, MT, Moore, PJ, Smale, DA, Alexander, LV, Benthuysen, JA, Feng, M, Gupta, AS, Hobday, AJ and Holbrook, NJ, (2018). 'Longer and more frequent marine heatwaves over the past century'. *Nature communications*, vol 9, p. 1324. doi:10.1038/s41467-018-03732-9.

Pearce, AF and Feng, M (2013). 'The rise and fall of the "marine heat wave" off Western Australia during the summer of 2010/2011'. *Journal of Marine Systems*, vol 111, pp. 139–156.

Pearce, A, Lenanton, R, Jackson, G, Moore, J, Feng, M and Gaughan, D (2011). 'The "marine heat wave" off Western Australia during the summer of 2010/11'. Fisheries Research Report No. 222, Department of Fisheries, Perth, Western Australia.

Poloczanska, ES, Brown, CJ, Sydeman, WJ, Kiessling, W, Schoeman, DS, Moore, PJ, Brander, K, Bruno, JF, Buckley, LB, Burrows, MT and Duarte, CM (2013). 'Global imprint of climate change on marine life'. *Nature Climate Change*, vol 3, no 10, pp. 919–925.

Power, AJ, Walker, RL, Payne, K and Hurley, D (2004). 'First occurrence of the nonindigenous green mussel, *Perna viridis* (Linnaeus, 1758) in coastal Georgia, United States'. *Journal of Shellfish Research*, vol 23, no 3, pp. 741–745.

Robinson, L, Hobday, AJ, Possingham, HP and Richardson, AJ (2015). 'Trailing edges projected to move faster than leading edges for large pelagic fish under climate change'. *Deep Sea Research II*, vol 113, pp. 225–234.

Smale, DA and Wernberg, T (2012). 'Ecological observations associated with an anomalous warming event at the Houtman Abrolhos Islands, Western Australia'. *Coral Reefs*, vol 31, no 2, p. 441.

Smale, DA and Wernberg, T (2013). 'Extreme climatic event drives range contraction of a habitat-forming species'. *Proceedings of the Royal Society B: Biological Sciences*, vol 280, p. 1754. doi:10.1098/rspb.2012.2829.

Smith, KA, Chambers, N and Jackson, G, (2011). 'Unusual biological events coinciding with warm ocean conditions along the south-west coast of WA', in AF Pearce, R Lenanton, G Jackson, J Moore, M Feng and D Gaughan (eds.), 'The "marine heat wave" off Western Australia during the summer of 2010/2011'. Department of Fisheries Research Report No. 222. Department of Fisheries, Perth, Western Australia.

Spalding, M, Burke, L, Wood, SA, Ashpole, J, Hutchison, J and zu Ermgassen, P (2017). 'Mapping the global value and distribution of coral reef tourism'. *Marine Policy*, vol 82, pp. 104–113.

Thompson, PA, Bonham, P, Thomson, PG, Rochester, W, Doblin, MA, Waite, AM, Richardson, A and Rousseaux, CS (2015). 'Climate variability drives plankton community composition changes: The 2010–2011 El Niño to La Niña transition around Australia'. *Journal of Plankton Research*, vol 37, pp. 966–984.

Thomson, DP, Bearham D, Graham F and Eagle JV (2011). 'High latitude, deeper water coral bleaching at Rottnest Island, Western Australia'. *Coral Reefs*, vol 30, p. 1107. doi:10.1007/s00338-011-0811-x.

Thomson, JA, Burkholder, DA, Heithaus, MR, Fourqurean, JW, Fraser, MW, Statton, J and Kendrick, GA (2015). 'Extreme temperatures, foundation species, and abrupt ecosystem change: An example from an iconic seagrass ecosystem'. *Global Change Biology*, vol 21, no 4, pp. 1463–1474.

Viejo, RM, Martínez, B, Arrontes, J, Astudillo, C and Hernández, L (2010). 'Reproductive patterns in central and marginal populations of a large brown seaweed: Drastic changes at the southern range limit'. *Ecography*, vol 34, pp. 75–84.

Wernberg, T, Smale, DA, Tuya, F, Thomsen, MS, Langlois, TJ, De Bettignies, T, Bennett, S and Rousseaux, CS (2012). 'An extreme climatic event alters marine ecosystem structure in a global biodiversity hotspot'. *Nature Climate Change*, vol 3, no 1, pp. 78–82.

Wernberg, T, Bennett, S, Babcock, RC, de Bettignies, T, Cure, K, Depczynski, M, Dufois, F, Fromont, J, Fulton, CJ, Hovey, RK and Harvey, ES (2016). 'Climate-driven regime shift of a temperate marine ecosystem'. *Science*, vol 353, no 6295, pp. 169–172.

Wu, L, Cai, W, Zhang, L, Nakamura, H, Timmermann, A, Joyce, T, McPhaden, M, Alexander, MA, Qiu, B, Visbeck, M, Chang, P and Giese, B (2012). 'Enhanced warming over the global subtropical western boundary currents'. *Nature Climate Change*, vol 2, pp. 161–166. doi:110.1038/NCLIMATE1353.

Zhang N, Feng M, Hendon HH, Hobday AJ and Zinke J (2017). 'Opposite polarities of ENSO drive distinct patterns of coral bleaching potentials in the southeast Indian Ocean'. *Scientific Reports*, vol 7, p. 2443. doi:10.1038/s41598-017-02688-y.

9 Local, community-led interventions to address global-scale problems and environmental extremes in coastal ecosystems

Christopher D. Hepburn

Introduction

The Anthropocene has fundamentally modified marine ecosystems at a global scale. Hundreds of years of local degradation have resulted in reductions in ecosystem function that are now detectable at global scales (Parmesan and Yohe 2003; Halpern *et al.* 2008; Estes *et al.* 2011). After a period of unprecedented and ongoing local change, broader global and regional changes in the environment are occurring due to the impacts of increasing carbon dioxide (CO_2) concentrations (Royal Society 2005; Vermeer and Rahmstorf 2009; Lyman *et al.* 2010). Ocean warming, sea level rise and changing ocean chemistry have already been detected (Doney *et al.* 2012). Other changes such as increased storm frequency have also been attributed to a changing atmosphere and ocean climate (Coumou and Rahmstorf 2012). Our oceans are predicted to become warmer, more acidic, more stratified and be subjected to more extreme events such as storms and marine heatwaves (Boyd and Law 2011; Gruber 2011; Doney *et al.* 2012) (Marine heatwaves are explored in Chapter 8 which highlights the implications of change.) These changes are significant and represent a major adjustment, even a phase shift, for how our oceans will function in a high CO_2 world. Modification of physical, chemical and physiological processes are altering the functioning of marine ecosystems and the values they provide (Brander 2007; McCauley *et al.* 2015).

Coastal seas, the sensitive interface between the land and open ocean, have perhaps been most affected by local environmental and habitat degradation due to human activities over the last 150–300 years (Lotze *et al.* 2006). Today few, if any, coastal seas remain in an unmodified state. Iconic and globally important kelp forest (Johnson *et al.* 2011; Filbee-Dexter and Wernberg 2018), seagrass (Waycott *et al.* 2009) and coral reef (Hoegh-Guldberg *et al.* 2007) ecosystems are in decline. Productive coastal areas provide the majority of global fisheries, are important for extraction of oil and natural gas and their biogenic habitats are important in the reducing impacts of extreme events such as erosion due to storms (Barbier *et al.* 2011). Pollution (Foster and

Schiel 2010), invasive species (Molnar *et al.* 2008), over-fishing (Pauly 1998; Armstrong and Falk-Petersen 2008; Watson *et al.* 2012) and coastal development (Bulleri and Chapman 2010) are local stressors that have all degraded the values that coastal marine ecosystems provide (Worm *et al.* 2006).

The changing environment driven by increasing atmospheric CO_2 will modulate the effects of local stressors in coastal ecosystems (Harley *et al.* 2006; Russell *et al.* 2009). Local stressors exert strong, often direct, influences on ecosystem functioning (Airoldi 1998; Crain *et al.* 2008). Extreme events such as marine heatwaves, floods and storms are often attributed to changing climate but to assign direct causation to global change is difficult. High levels of environmental variability in coastal environments also make subtler, long-term patterns of change difficult to detect. A good example of this is ocean acidification (OA), the changing carbon chemistry of the oceans' surface as a direct result of anthropogenic derived CO_2 being absorbed by the ocean (Royal Society 2005; Doney *et al.* 2009). OA has been detected at a range of sites in oceanic waters over the last 30 years; anthropogenic CO_2 emissions have reduced the pH of oceanic water by 0.0013–0.0026 units over two or three decades (Bates *et al.* 2014). Ocean pH is estimated to have decreased by 0.1 (equivalent to a 30 per cent increase in H+ concentration) since the industrial revolution (Turley *et al.* 2006) and will fall a further 0.33 units by 2100 (Law *et al.* 2018). However, in coastal seawater pH can vary by as much 0.94 of a pH unit during a single day due to photosynthesis (Cornwall *et al.* 2013). Other factors such as temperature, salinity, nutrient loading and interactions with sediment can also strongly modify pH in coastal waters (Law *et al.* 2018). These factors make detecting the gradual change in pH due to OA, set against an extremely variable signal in coastal seawater, intractable. Trying to detect, understand and predict how slow, subtle, yet probably significant changes in global ocean pH will modify coastal seawater, and how these changes will impact on complex coastal ecosystems, is particularly challenging.

Despite the complexity and variability for coastal ecosystems, a body of literature has foretold a dire future. For example, the changing environment will damage and even eliminate key biogenic habitats in many areas. Species ranges will change, and competitive interactions between species will be modified (e.g. Hepburn *et al.* 2011; Johnson *et al.* 2011; Doney *et al.* 2012; Wernberg *et al.* 2012). Much of the news is bad and for those struggling to manage coastal environments, ecosystems and resources, facing the problems of global change is an apparently impossible task.

New research looks for positive ways of addressing change (e.g. Parker *et al.* 2010; Hurd 2015) but the problems are overwhelming in scale and we have limited capability to manipulate marine ecosystems. Local people who manage or use marine ecosystems quite often jokingly say, 'If it is all so hopeless why not maximise the profits of today off these doomed ecosystems? Let us harvest and mine and modify – let us make the most of today.' What does one say to an argument of this type when so much of the data and associated commentary predicts such a sad future for coastal ecosystems?

In this chapter the story is presented of a community group with responsibility for the management of connected coastal ecosystems and fisheries across freshwater, estuarine and marine boundaries. For the community managers involved, an understanding of the impacts of global scale environmental change and extreme events on local ecosystems is an important priority, with the wellbeing of generations to come intrinsic to their management role.

Local management of a fisheries ecosystem in a fishery Customary Protection Area

Place-based management and more local approaches have been suggested to be a key component of the revolution in management of marine fisheries to more holistic, ecosystem focussed approaches (Pauly 1997; Berkes 2012). Taiāpure is a type of fishery Customary Protection Area (CPA) that provides a legislative platform which allows Māori (New Zealand's indigenous people), Iwi (tribe) and hapū (sub-tribe) to manage local fisheries of local and sustained importance (Bess and Rallapudi 2007; Hepburn *et al.* 2010; Jackson *et al.* 2010). Taiāpure were established under the 'interim' Fisheries Settlement in 1989 to partially address breaches of Article Two of the Treaty of Waitangi, New Zealand's founding document, by the Crown against the rights of Iwi/hapū over fisheries (Jackson 2013).

The East Otago Taiāpure provides a 25-year case study of community driven management of a local fishery ecosystem along a 25 km stretch of coastline on New Zealand's South Island (see Figure 9.1). The East Otago Taiāpure is within the rohe moana (marine tribal area) of Ngāi Tahu (tribe), who possess rights to over 250,000 km^2 of New Zealand's territorial sea and exclusive economic zone. It is the largest fisheries area of any Iwi in New Zealand. The Taiāpure falls under the authority of the local hapū, Kāti Huirapa ki Puketeraki, and is supported by Te Rūnanga o Ngāi Tahu (Ngāi Tahu's tribal council).

Key in the establishment of the Taiāpure were concerns surrounding the sustained decline of the Pāua (*Haliotis iris*, abalone) fishery (Figures 9.2 and 9.3) (Jackson 2011). Pāua is a cultural keystone and an important source of food, as well as for traditional practices and ways of life (Garibaldi and Turner 2004; Turner *et al.* 2013). Interviews with over 100 individuals who have a direct relationship with the marine environment and resources therein have shown that pāua is the number one species of concern and importance for southern New Zealanders (McCarthy *et al.* 2014). Abalone fisheries globally offer a sad tale of overexploitation and collapse (Prince 2005; Rogers-Bennett *et al.* 2013). In the Taiāpure many of the local management interventions (bag limit reduction, spatial closures) have focussed on pāua (Gnanalingam and Hepburn 2015). The Taiāpure management committee have also fought hard to protect the kelp forest that provides habitat and food for pāua, many other species and wider values (see Figures 9.5, 9.6 and 9.7). Tangata tiaki (legislatively empowered customary fishery managers) and members of the Taiāpure

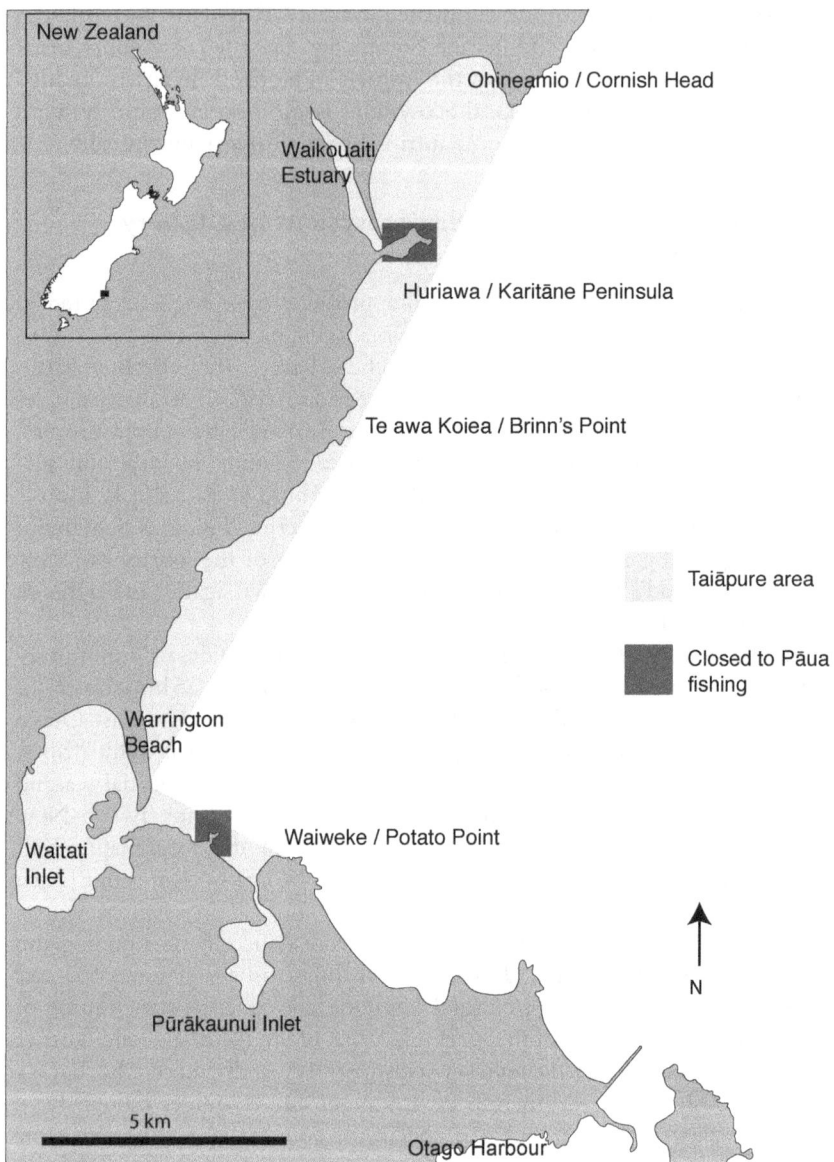

Figure 9.1 The East Otago Taiāpure.

Note

The Taiāpure extends approximately 25 km along the coast and includes four estuaries, a number of beaches and headlands and extensive subtidal reef and soft sediment habitats. The Taiāpure sits next to and contains many significant historical and mahinga kai (areas for wild food gathering) sites.

Figure 9.2 A pāua (blackfoot abalone, *Haliotis iris*) in a deeper kelp forest habitat in the East Otago Taiāpure.

Source: Christopher D. Hepburn.

Figure 9.3 A group of pāua in shallow wave exposed habitat in a bull kelp (*Durvillaea* spp.) bed.

Source: Christopher D. Hepburn.

management committee use kaitiakitanga (imperfectly translated as 'environ-mental guardianship by tangata whenua' – people of the land), to protect and restore the values of the local fishery in a holistic manner (Hepburn *et al.* 2010). Taiāpure managers use the best available information, often local knowledge and science, to make decisions. Mātauranga (traditional ecological knowledge related to the natural world in this context) is key in the manage-ment of the Taiāpure. In the Taiāpure, mātauranga guides science, providing a focus for fishery monitoring and supporting the development of research questions (Figure 9.4).

Local problems swamp global scale issues for local managers

The pāua population and kelp forests of the Taiāpure are sensitive to local stressors and predicted to be further impacted by the changing environment and extreme events. In East Otago, kelp forests are formed by *Macrocystis pyrifera*, the giant kelp (Figure 9.5), an autogenic ecosystems engineer (an organism that changes the environment with its own physical structure) (Jones and Lawton 1994). In shallower water other macroalgal species play similar roles; the massive bull kelps *Durvillaea antarctica* and *Durvillaea willana* are particularly important (Figure 9.6). A diversity of other more diminutive macroalgal species are also found in the many layered structures of these

Figure 9.4 Wading surveys are conducted in the rāhui (a closed fishery area in this case) at Huriawa Peninsula.

Source: Christopher D. Hepburn.

Note
Shallow sections of the reef are of primary value in a customary fishery as pāua gatherers have traditionally never used a mask and snorkel but gather at low tide.

Figure 9.5 A *Macrocystis pyrifera* kelp forest at Huriawa Peninsula, East Otago Taiāpure.
Source: Christopher D. Hepburn.
Note
This area has been a key site for fisheries, habitat and climate change research.

undersea forests (Hepburn *et al.* 2011) (Figure 9.7). Kelp forests provide a key habitat for reef fishes (Win 2011) and their canopy provides a surface for attachment for many organisms (Hepburn and Hurd 2005), not unlike the canopy of a tropical rainforest. Kelp forests are key kōhanga (nursery) areas for fish (Win 2011) and are thought to allow for the settlement of larvae of koura (crayfish, *Jasus edwardsii*) and perhaps pāua by slowing water flow, allowing passing larvae to reach the seabed and providing a refuge for newly settled recruits (Hinojosa *et al.* 2015; Hesse *et al.* 2016). Kelp forests can dampen waves and as a result may help prevent coastal erosion (Gaylord *et al.* 2012; Smale *et al.* 2013) (the ecosystem services that natural systems can provide to mitigate flooding and erosion are examined further in Chapter 5).

The extensive offshore *Macrocystis* kelp forests (Fyfe *et al.* 1999) within and surrounding the Taiāpure are probably the most extensive and important example of this type of marine ecosystem in New Zealand. Kelp forests are sensitive to local stressors, such as sedimentation and nitrogen loading (Airoldi 1998; Gorman *et al.* 2009; Foster and Schiel 2010). Local research has demonstrated that kelp forests in the Taiāpure are light limited for much of the year and that sediment run-off from the land can reduce their productivity and extent (Hepburn *et al.* 2011; Pritchard *et al.* 2013; Desmond *et al.* 2015). The invasive *Undaria pinnatifida* has reached the Taiāpure from a nearby port

Figure 9.6 Bull kelp (*Durvillaea antarctica* and *Durvillaea willana*) on the southern more wave-exposed coast of Huriawa.

Source: Christopher D. Hepburn.

Note
Bull kelp provides important habitat and food for pāua and other species.

(Russell *et al.* 2007) and research suggests that it provides less value to local ecosystems than native kelps (Suárez-Jiménez *et al.* 2015, 2017). Research focussed on pāua shows a declining fishery (Gnanalingam and Hepburn 2015) and impacts of sedimentation on juvenile habitat and juvenile behaviour (Chew *et al.* 2013).

The Taiāpure has been the site for research on how kelp forests and associated fisheries will respond to changing coastal environments for more than ten years. This research has been strongly supported by the managers of the Taiāpure and the findings have been shared extensively with the local community. Much of this research has focussed on calcifying coralline algae that cover reef surfaces and fulfil many roles (Nelson 2009). They are the cold-temperate version of a coral reef. OA will impact negatively on the growth of coralline communities through lowered pH and perhaps benefit non-calcifying macroalgae competitors by providing additional CO_2 for photosynthesis. Differential responses of different groups of habitat forming species could result in shifts in habitat structure with flow-on effects for fisheries (Hepburn *et al.* 2011). Perhaps most interestingly was the finding that

Figure 9.7 Coralline algae beneath a canopy of macroalage (*Xiphophora gladiata*) in a shallow section of kelp forests at Huriawa.

Source: Christopher D. Hepburn.

Note
Coralline algae provide critical habitat to support fisheries in the Taiāpure and are sensitive to ocean acidification.

photosynthesis of kelp (Cornwall, *et al.* 2013) and the physical structure of macroalgal canopy (Cornwall *et al.* 2015) can allow for increased pH in areas where sensitive calcifying organisms live (Cornwall *et al.* 2014). This may provide benefits for calcification and a refuge from OA (Hurd 2015).

Pāua are calcifying molluscs, a group that has been identified as being very sensitive to changing ocean pH through meta-analysis of OA research globally (Kroeker *et al.* 2010). Early life stages appear particularly sensitive and laboratory experiments show that OA can reduce the growth, shell formation and survivorship of juveniles (Cunningham *et al.* 2016). Coralline algae play key roles in facilitating the settlement and metamorphosis of planktonic pāua larvae (Naylor *et al.* 2006), support growth of early life stages, and are also one of the most OA sensitive groups of marine organisms based on local (see above) and international research (Kroeker *et al.* 2010).

The research at the Taiāpure continues as a long-term monitoring site for light, pH and many other environmental variables established in 2006 at Te Awa a Mokihi (Butterfly Bay), Huriawa. Records from the nearby Portobello Marine Laboratory show that annual sea surface temperature (SST) has increased more than half a degree, on average, over the 65-year time series, and recent satellite data suggests a general pattern of warming focussed on southern New Zealand (Shears and Bowen 2017). The summer of 2017–2018 was the hottest in living memory, and recorded in the 65-year temperature Portobello data set. Records also suggest that New Zealand was subjected to a heatwave focussed in southern regions. Kelp forests are sensitive to warming oceans and extreme events; for example, strengthening of the East Australian Current and a warming of water has all but eliminated *M. pyrifera* from Tasmania's East Coast (Johnson *et al.* 2011). Summer storms and floods appear to be more common and more extreme in the area of the Taiāpure and the sediment released could have implications for already light limited kelp forests (Hepburn *et al.* 2011; Pritchard *et al.* 2013; Desmond *et al.* 2015).

Currently, the managers of the East Otago Taiāpure face more direct threats to the sustainability of local fisheries through locally driven stressors than through stressors driven by increases in atmospheric CO_2. Dealing with issues such as sedimentation, overfishing, water extraction, eutrophication, port dredging and harvests of habitat forming kelp are difficult but possible in the current political climate. For example, the Taiāpure Management Committee has acted through its mandate to propose regulations to the Minister of Fisheries to alter customary, commercial and/or recreational fishing for the conservation and management of fish, aquatic life or seaweed (Fisheries Act 1996). Local regulations have significantly reduced recreational bag limits for pāua, finfish and a range of shellfish species (Hepburn *et al.* 2010) and established closed areas/rāhui (Gnanalingam and Hepburn 2015). Pāua fishery restocking using juveniles from aquaculture juveniles has also been facilitated by the Taiāpure Management Committee (Gillies 2014). Agreements have also been reached with commercial pāua fishers who have pledged to stop harvesting in the Taiāpure.

Despite these interventions monitoring of the pāua population shows a continued decline in the fishery and only closed areas are increasing or at worst remaining steady (Hepburn *et al.* 2016). These issues are likely associated with the ineffectiveness of individual bag limits, illegal take, recruitment limitation or habitat/environmental degradation (Gnanalingam and Hepburn 2015). A proposal for a closure on pāua fishing for the entire Taiāpure to allow stocks to recover has recently been lodged with the Minister of Fisheries. This work occurs alongside many other projects that the Taiāpure Management Committee are involved in. It is likely that these interventions are improving fisheries and the environment, but assessing success or failure at the present time is very difficult. Detecting the benefits of community-led ecosystem management and restoration applied across ecosystems (Hepburn *et al.* 2010) is a significant scientific challenge. A lack of long-term time series for key species and the environment and the fact that some information is currently too difficult or costly to acquire also makes such assessments very challenging at present.

Tentative observations of local managers suggest that a changing climate is beginning to impact on local ecosystems. These changes will interact with local stressors (see below), so that predicting what may happen is difficult. Taiāpure managers are concerned about climate change but dealing with the drivers in the current political climate is intractable for them or even the New Zealand government. The Taiāpure Management Committee has naturally developed a strategy to address the changing climate, which is applied to most problems they face. The local community manager learns by supporting research, being observant and keeping records of personal interactions with the fishery, ecosystems and the environment. The local community is working to improve the environment through replanting, fishery restocking and the monitoring of rivers and catchments and the coastal sea to inform management and drive change. The goal of the East Otago Taiāpure is to achieve broadly sustainable fisheries within a customary fishery context. To achieve this, fishing practices and the people who maintain traditions through fishing must also be supported by management initiatives. Because of this the Taiāpure Management Committee has taken a precautionary approach, reducing catch through bag limit reductions, closed areas, removing commercial pāua harvest and a recent proposal to close the entire Taiāpure to pāua fishing. These interventions are designed to address the sustained decline of local stocks and will hopefully lead to the maintenance of local fisheries and traditions now and in the future. These approaches will also provide resilience for local ecosystems, associated fisheries and the practices and ways of life of local people in a changing and more extreme climate.

Managing coastal ecosystems in a changing and more extreme environment

In New Zealand, as in many places, the last 200 years has been a period of unprecedented change. The development of intensive agriculture, urbanisation, fishing and shipping has profoundly changed coastal ecosystems. Are the changes predicted to occur over the next 200 years as extreme? They may not be. However, change never occurs in isolation or without consideration of historical context. Our natural ecosystems today are not as resilient they might once have been, and the effects of climate change will be cumulative and provide further pressures on already crumbling foundations. Additive, antagonistic and synergistic interactions between local and global stressors makes the story complex and predictions difficult (Crain *et al.* 2008; Smale *et al.* 2013).

Yet, this knowledge also suggests some hope. Local restoration and more effective management of coastal ecosystems could provide a pathway for reducing the impacts of global change on systems through locally focussed interventions. Dealing with local stressors, overexploitation and poor management allows local people, powerless to bring about change to deal with the global drivers of increasing CO_2, to find a way to maintain ways of life associated with productive and functioning coastal ecosystems. Short term extremes such as heatwaves, flooding and storm disturbance can most certainly be dealt with better by diverse and intact ecosystems. Ecosystem resilience is provided by functional redundancy, where many species fulfil the same role in an ecosystem (e.g. Steneck and Graham 2002; Micheli and Halpern 2005), buffering, where one group of species modifies the local environment (e.g. Cornwall *et al.* 2013), and the high reproductive potential of unexploited stocks (e.g. Pauly 1997). These mechanisms provide ecosystem resilience and stability and allow ecosystems to respond to and resist disturbance and mortality due to extreme events. Functional redundancy, buffering capacity and reproductive potential have all been eroded by past habitat and environmental degradation and exploitation. Exploitation that is sustainable today may not be sustainable in a high CO_2 world. For example, fisheries cannot absorb the fishing pressure they once could (Worm *et al.* 2006; Armstrong and Falk-Petersen 2008). Fishing reduces the age, size, and geographic diversity of populations as well as the biodiversity of marine ecosystems, thus reducing their ability to withstand fishing pressure (Worm *et al.* 2006) and respond to the changing environment (Brander 2007). Fisheries management must adjust to deal with environmental change and apply a more precautionary approach (Botsford *et al.* 1997; Brander 2007).

Conclusion

Management that empowers local people with a long-term interest in an ecosystem will be key in addressing environmental change and in a more extreme future climate. Management models with people at the centre of the ecosystem, as is often the case with approaches used by indigenous people, may be flexible and responsive enough to deal with and prepare for future change. These approaches certainly reflect the central position that people now have in coastal ecosystems and human impacts will only strengthen in the future. Protection, restoration and gradual improvement of coastal environments and ecosystems through local approaches across coastal ecosystems and connected habitats (e.g. wetlands, coastal forests and alpine water sheds) are of greatest value when preparing for extreme events in the future (e.g. storms, flooding, heatwaves). Developing restoration strategies and tools for species that may broadly influence ecosystem functioning and facilitate habitats for other species such as kelp, seagrass and coral will provide greatest value in dealing with change in coastal ecosystems while providing added benefits such as coastal protection (see Chapter 5). Identification and development of strains of species resistant to environmental extremes and change could be the last hope when natural populations are pushed beyond their physiological limits.

However, we must not wait for the engineering of biotechnological panaceas that will save our coastal seas from the problems we face. Such interventions are expensive (Bayraktarov *et al.* 2016) and even if effective are unlikely to be able to address issues that are on such a large scale as we deal with global change. The development of new tools to address coming change must occur in concert with the less glamorous work of local management and restoration.

The long-term, intergenerational view of East Otago Taiāpure is not unusual for local managers, particularly for management by indigenous communities. The 200-year timetable for ecosystem and community restoration set in motion by leaders of the Taiāpure means that global scale environmental change is a major factor that must be considered. For today, the best approach is to build understanding of what will happen and use this knowledge to focus interventions on those that offer the best potential of providing the resilience of social–ecological systems against change. Building community capacity and predictive knowledge and acting to address local stressors will benefit local communities today, but also provide value for the generations to come who face a very uncertain future.

Acknowledgements

This chapter is based on a presentation entitled, 'Bracing once more for change: Is science helping local people prepare for climate change impacts on coastal ecosystems?', at the Matariki Network of Universities workshop 'Oceans and Blue Economy' in December 2017. I acknowledge the contribution of the co-authors of the paper presented at the workshop, Brendan Flack, Nigel Scott and Daniel Pritchard. Members of the East Otago Taiāpure Management Committee and the Mahinga Kai Monitoring and Enhancement team from Te Ao Turoa, Te Rūnanga o Ngāi Tahu were key in formulating this chapter through discussions at monthly meetings and their work over the last ten years. I also acknowledge the many University of Otago students and staff who have worked in the Taiāpure and collaborators and mentors, particularly Derek Richards, Catriona Hurd, Kim Currie, Henrik Moller and Anne-Marie Jackson. This work was conducted with Te Tiaki Mahinga Kai, a partnership programme between local communities and researchers (www.mahingakai.org.nz).

References

Airoldi, L. (1998). 'Roles of disturbance, sediment stress, and substratum retention on spatial dominance in algal turf'. *Ecology*, vol 79, no 8, pp. 2759–2770.

Armstrong, C.W. and Falk-Petersen, J. (2008). 'Habitat-fisheries interactions: A missing link?'. *ICES Journal of Marine Science*, vol 65, no 6, pp. 817–821.

Barbier, E.B., Hacker, S.D., Kennedy, C., Koch, E.W., Stier, A.C. and Silliman, B.R. (2011). 'The value of estuarine and coastal ecosystem services'. *Ecological Monographs*, vol 81, no 2, pp. 169–193.

Bates N., Astor Y., Church M., Currie K., Dore J., Gonaález-Dávila M., Lorenzoni, L., Muller-Karger, F., Olafsson, J. and Santa-Casiano M. (2014). 'A time-series view of changing ocean chemistry due to ocean uptake of anthropogenic CO2 and ocean acidification'. *Oceanography*, vol 27, no 1, pp. 126–141.

Bayraktarov, E., Saunders, M.I., Abdullah, S., Mills, M., Beher, J., Possingham, H.P., Mumby, P.J. and Lovelock, C.E. (2016). 'The cost and feasibility of marine coastal restoration'. *Ecological Applications*, vol 26, no 4, pp. 1055–1074.

Berkes, F. (2012). 'Implementing ecosystem-based management: Evolution or revolution?'. *Fish and Fisheries*, vol 13, no 4, pp. 465–476.

Bess, R. and Rallapudi, R. (2007). 'Spatial conflicts in New Zealand fisheries: The rights of fishers and protection of the marine environment'. *Marine Policy*, vol 31, no 6, pp. 719–729.

Botsford, L.W., Castilla, J.C. and Peterson, C.H. (1997). 'The management of fisheries and marine ecosystems'. *Science*, vol 277, no 5325, pp. 509–515.

Boyd, P.W. and Law, C.S. (2011) 'An ocean climate change atlas for New Zealand waters'. NIWA Information Series No. 79, National Institute of Water and Atmospheric Research, Wellington, New Zealand.

Brander, K.M. (2007). 'Climate change and food security special feature: Global fish production and climate change'. *Proceedings of the National Academy of Sciences*, vol 104, no 50, pp. 19709–19714.

Bulleri, F. and Chapman, M.G. (2010). 'The introduction of coastal infrastructure as a driver of change in marine environments'. *Journal of Applied Ecology*, vol 47, no 1, pp. 26–35.

Chew, C.A., Hepburn, C.D. and Stephenson, W. (2013). 'Low-level sedimentation modifies behaviour in juvenile *Haliotis iris* and may affect their vulnerability to predation'. *Marine Biology*, vol 160, no 5, pp. 1213–1221.

Cornwall, C.E., Hepburn, C.D., McGraw C.M., Currie K.I, Pilditch, C, Hunter K.A., Boyd P.W. and Hurd C.L. (2013). 'Diurnal fluctuations in seawater pH influence the response of a calcifying macroalgae to ocean acidification'. *Proceedings of the Royal Society B: Biological Sciences*, vol 280, no 1772, p. 20132201.

Cornwall, C.E., Boyd P.W., McGraw C.M., Hepburn, C.D., Pilditch, C.A, Morris J.N., Smith A.M. and Hurd C.L. (2014). 'Diffusion boundary layers ameliorate the negative effects of ocean acidification on the temperate coralline macroalga *Arthrocardia corymbosa*'. *PLoS ONE*, vol 9, no 5, p. e97235.

Cornwall, C.E., Pilditch, C., Hepburn, C.D. and Hurd C.L. (2015). 'Canopy macroalgae influence understorey corallines' metabolic control of near-surface pH and oxygen concentration'. *Marine Ecology Progress Series*, vol 525, pp. 81–95.

Coumou, D. and Rahmstorf, S. (2012). 'A decade of weather extremes'. *Nature Climate Change*, vol 2, pp. 491–496.

Crain, C.M., Kroeker, K. and Halpern, B.S. (2008). 'Interactive and cumulative effects of multiple human stressors in marine systems'. *Ecology Letters*, vol 11, pp. 1304–1315.

Cunningham, S.C., Smith, A.M. and Lamare, M.D. (2016). 'The effects of elevated pCO_2 on growth, shell production and metabolism of cultured juvenile abalone, *Haliotis iris*'. *Aquaculture Research*, vol 47, pp. 2375–2392.

Desmond, M.J., Pritchard, D.W. and Hepburn, C.D. (2015). 'Light limitation within southern New Zealand kelp forest communities'. *PLoS ONE*, vol 10, no 4, p. e0123676.

Doney, S.C., Fabry, V.J., Feely, R.A. and Kleypas, J.A. (2009). 'Ocean acidification: The other CO_2 problem'. *Annual Review of Marine Science*, vol 1, no 1, pp. 169–192.

Doney, S.C., Ruckelshaus, M., Emmett Duffy, J., Barry, J.P., Chan, F., English, C.A., Galindo, H.M., Grebmeier, J.M., Hollowed, A.B., Knowlton, N., Polovina, J., Rabalais, N.N., Sydeman, W.J. and Tally, L.D. (2012). 'Climate change impacts on marine ecosystems'. *Annual Review of Marine Science*, vol 4, no 1, pp. 11–37.

Estes, J.A., Terborgh, J., Brashares, J.S., Power, M.E., Berger, J., Bond, W.J., Carpenter, S.R., Essington, T.E., Holt, R.D., Jackson, J.B.C., Marquis, R.J.,

Oksanen L., Oksanen, T., Paine R.T., Pikitch, E.K., Ripple, W.J., Sandin, S.A., Scheffer, M., Schoener, T.W., Shurin, J.B., Sinclair, A.R.E., Soulé, M.E, Virtanen, R. and Wardle D.A. (2011). 'Trophic downgrading of planet earth'. *Science*, vol 333, no 6040, pp. 301–306.

Filbee-Dexter, K. and Wernberg, T. (2018). 'Rise of turfs: A new battlefront for globally declining kelp forests'. *BioScience*, vol 68, no 2, pp. 64–76.

Fisheries Act (1996). New Zealand government. Available at www.legislation.govt. nz/act/public/1996/0088/latest/DLM394192.html.

Foster, M.S. and Schiel, D.R. (2010). 'Loss of predators and the collapse of southern California kelp forests (?): Alternatives, explanations and generalizations'. *Journal of Experimental Marine Biology and Ecology*, vol 393, no 1–2, pp. 59–70.

Fyfe, J., Israel, S.A., Chong, A., Ismail, N., Hurd, C.L. and Probert K. (1999). 'Mapping marine habitats in Otago, southern New Zealand'. *Geocarto International*, vol 14, no 3, pp. 17–28.

Garibaldi, A. and Turner, N. (2004). 'Cultural keystone species: Implications for ecological conservation and restoration'. *Ecology and Society*, vol 9, no 3, p. 1.

Gaylord, B., Nickols, K.J. and Jurgens, L. (2012). 'Roles of transport and mixing processes in kelp forest ecology'. *Journal of Experimental Biology*, vol 215, no 6, pp. 997–1007.

Gillies, T.T. (2014). 'Reseeding of *Haliotis iris* in a customary fisheries context'. MSc thesis, University of Otago, New Zealand.

Gnanalingam, G. and Hepburn, C. (2015). 'Flexibility in temporary fisheries closure legislation is required to maximise success'. *Marine Policy*, vol 61, pp. 39–45.

Gorman, D., Russell, B.D. and Connell, S.D. (2009). 'Land-to-sea connectivity: Linking human-derived terrestrial subsidies to subtidal habitat change on open rocky coasts'. *Ecological Applications*, vol 19, no 5, pp. 1114–1126.

Gruber, N. (2011). 'Warming up, turning sour, losing breath: Ocean biogeochemistry under global change'. *Philosophical Transactions of the Royal Society A: Mathematical, Physical and Engineering Sciences*, vol 369, no 1943, pp. 1980–1996.

Halpern, B.S., Walbridge, S., Selkoe, K.A., Kappel, C.V., Micheli, F., D'Agrosa, C., Bruno, J.F., Casey, K.S., Ebert, C., Fox, H.E., Fujita, R., Heinemann, D., Lenihan, H.S., Madin, E.M.P., Perry, M.T., Selig, E.R., Spalding, M., Steneck, R. and Watson, R. (2008). 'A global map of human impact on marine ecosystems'. *Science*, vol 319, no 5865, pp. 948–952.

Harley, C.D.G., Randall Hughes, A., Hultgren, K.M., Miner, B.G., Sorte, C.J.B., Thornber, C.S., Rodriguez, L.F., Tomanek, L. and Williams, S.L. (2006). 'The impacts of climate change in coastal marine systems'. *Ecology Letters*, vol 9, no 2, pp. 228–241.

Hepburn, C.D. and Hurd, C.L. (2005). 'Conditional mutualism between the giant kelp *Macrocystis pyrifera* and colonial epifauna'. *Marine Ecology Progress Series*, vol 302, pp. 37–48.

Hepburn, C.D., Jackson, A.M., Vanderburg, P.H., Kainamu A. and Flack, B. (2010). 'Ki uta ki tai: From the mountains to the sea. Holistic approaches to customary fisheries management'. Proceedings of the 4th International Indigenous Conference on Traditional Knowledge: Kei muri i te kāpara he tangata, Recognizing, engaging, understanding difference, Auckland, 6–9 June 2010, pp. 142–148.

Hepburn, C.D., Pritchard, D.W., Cornwall, C.E., McLeod, R.J., Beardall, J.B., Raven, J.A. and Hurd, C.L. (2011). 'Diversity of carbon use strategies in a kelp forest community: Implications for a high CO2 ocean'. *Global Change Biology*, vol 17, no 7, pp. 2488–2497.

Hepburn, C.D., Subritzky, P., Richards, D.K. and Pritchard, D.W. (2016). 'Status of the East Otago Taiapure Pāua Fishery 2008–2016'. *He Kōhinga Rangahau*, University of Otago, Dunedin.

Hesse, J., Stanley, J.A. and Jeffs, A.G. (2016). 'Relative predation risk in two types of habitat for juvenile Australasian spiny lobsters, *Jasus edwardsii*'. *Marine Biology Research*, vol 12, no 9, pp. 895–906.

Hinojosa, I.A., Green, B.S., Gardner, C. and Jeffs, A. (2015). 'Settlement and early survival of southern rock lobster, *Jasus edwardsii*, under climate-driven decline of kelp habitats'. *ICES Journal of Marine Science*, vol 72, no 1, pp i59–i68.

Hoegh-Guldberg, O., Mumby, P.J., Hooten, A.J., Steneck, R.S., Greenfield, P., Gomez, E., Harvell, C.D., Sale, P.F., Edwards, A.J., Caldeira, K., Knowlton, N., Eakin, C.M., Iglesias-Prieto, R., Muthiga, N., Bradbury, R.H., Dubi A. and Hatziolos M.E. (2007). 'Coral reefs under rapid climate change and ocean acidification'. *Science*, vol 318, no 5857, pp. 1737–1742.

Hurd, C.L. (2015). 'Slow-flow habitats as refugia for coastal calcifiers from ocean acidification'. *Journal of Phycology*, vol 51, no 4, pp. 599–605.

Jackson, A.-M. (2011). 'Ki Uta Ki Tai: He Taoka Tuku Iho'. PhD thesis, University of Otago, New Zealand.

Jackson, A.-M. (2013). 'Erosion of Māori fishing rights in customary fisheries management'. *Waikato Law Review*, vol 21, pp. 59–75.

Jackson, A.-M., Hepburn, C.D. and East Otago Taiāpure Management Committee (2010). 'Rangatiratanga and customary fisheries management'. Proceedings of the 4th International Indigenous Conference on Traditional Knowledge: Kei muri i te kāpara he tangata, Recognizing, engaging, understanding difference, Auckland, 6–9 June 2010, pp. 165–170.

Johnson, C.R., Banks, S.C., Barrett, N.S., Cazassus, F., Dunstan, P.K., Edgar, G.J., Frusher, S.D., Gardner, C., Haddon, M., Helidoniotis, F., Hill, K.L., Holbrook, N.J., Hosie, G.W., Last, P.R., Ling, S.D., Melbourne-Thomas, J., Miller, K, Pecl G.T., Richardson, A.J., Ridgway, K.R., Rintoul, S.R., Ritz, D.A., Ross, D.J., Sanderson, J.C., Scoresby, S.A., Shepherd, A., Slotwinski, A., Swadling, K.M. and Taw, N. (2011). 'Climate change cascades: Shifts in oceanography, species' ranges and subtidal marine community dynamics in eastern Tasmania'. *Journal of Experimental Marine Biology and Ecology*, vol 400, no 1–2, pp. 17–32.

Jones, C. and Lawton, J. (1994). 'Organisms as ecosystem engineers'. *Oikos*, vol 69, no 3, pp. 373–386.

Kroeker, K.J., Kordas, R.L., Crim, R.N. and Singh, G.G. (2010). 'Meta-analysis reveals negative yet variable effects of ocean acidification on marine organisms'. *Ecology Letters*, vol 13, no 11, pp. 1419–1434.

Law, C., Bell, J., Bostock, H., Cornwall, C., Cummings, V., Currie, K., Davy, S., Gammon, M., Hepburn, C., Hurd, C., Lamare, M., Mikaloff-Fletcher, S., Nelson., W., Parsons, D., Ragg, N., Sewell, M., Smith, A. and Tracey, D. (2018). 'Ocean acidification in New Zealand waters: Trends and impacts'. *New Zealand Journal of Marine and Freshwater Research*, vol 52, no 2, pp. 155–195.

Lotze, H.K., Lenihan, H.S., Bourque, B.J., Bradbury, R.H., Cooke, R.G., Kay, M.C., Kidwell, S.M., Kirby, M.X., Peterson, C.H. and Jackson, J.B.C. (2006). 'Depletion, degradation, and recovery potential of estuaries and coastal seas'. *Science*, vol 312, no 5781, pp. 1806–1809.

Lyman, J.M., Good, S.A., Gouretski, V.V., Ishii, M., Johnson, G.C., Palmer, M.D., Smith, D.M. and Willis J.K. (2010). 'Robust warming of the global upper ocean'. *Nature*, vol 465, no 7296, pp. 334–337.

McCarthy A., Hepburn C.D., Scott N., Schweikert K., Turner, R. and Moller H. (2014). 'Local people see and care most? Severe depletion of inshore fisheries and its consequences for Māori communities'. *New Zealand Aquatic Conservation: Marine and Freshwater Ecosystems*, vol 24, pp. 369–390.

McCauley, D.J., Pinsky, M.L., Palumbi, S.R., Estes, J.A., Joyce, F.H. and Warner, R. R. (2015). 'Marine defaunation: Animal loss in the global ocean'. *Science*, vol 347, no 6219, p. 1255641.

Micheli, F. and Halpern, B.S. (2005). 'Low functional redundancy in coastal marine assemblages'. *Ecology Letters*, vol 8, no 4, pp. 391–400.

Molnar, J.L., Gamboa, R.L. and Revenga, C. (2008). 'Assessing the global threat of invasive species to marine biodiversity'. *Frontiers in Ecology and the Environment*, vol 6, no 9, pp. 485–492.

Naylor, R., Neill, K. and Stewart, R. (2006). 'Coralline algae and pāua settlement'. *Water & Atmosphere*, vol 14, no 2, pp. 14–15.

Nelson, W.A. (2009). 'Calcified macroalgae – critical to coastal ecosystems and vulnerable to change: A review'. *Marine and Freshwater Research*, vol 60, no 8, pp. 787–801.

Parker, L.M., Ross, P.M. and O'Connor, W.A. (2010). 'Populations of the Sydney rock oyster, *Saccostrea glomerata*, vary in response to ocean acidification'. *Marine Biology*, vol 158, no 3, pp. 689–697.

Parmesan, C. and Yohe, G. (2003). 'A globally coherent fingerprint of climate change impacts across natural systems'. *Nature*, vol 421, no 6918, p. 37.

Pauly, D. (1997). 'Putting fisheries management back in places'. *Reviews in Fish Biology and Fisheries*, vol 7, no 1, pp. 125–127.

Pauly, D. (1998). 'Fishing down marine food webs'. *Science*, vol 279, no 5352, pp. 860–863.

Prince, J. (2005). 'Combating the tyranny of scale for haliotids: Micro-management for microstocks'. *Bulletin of Marine Science*, vol 76, no 2, pp. 557–577.

Pritchard, D.W., Hurd, C.L., Beardall, J. and Hepburn, C.D. (2013). 'Survival in low light: Photosynthesis and growth of a red alga in relation to measured in situ irradiance'. *Journal of Phycology*, vol 49, no 5, pp. 867–879.

Rogers-Bennett, L., Hubbard, K.E. and Juhasz, C.I. (2013). 'Dramatic declines in red abalone populations after opening a "de facto" marine reserve to fishing: Testing temporal reserves'. *Biological Conservation*, vol 157, pp. 423–431.

Royal Society (2005). 'Ocean acidification due to increasing atmospheric carbon dioxide'. Policy document 12/05, The Royal Society, London.

Russell, B.D., Thompson, J.-A.I., Falkenberg, L.J. and Connell, S.D. (2009). 'Synergistic effects of climate change and local stressors: CO_2 and nutrient-driven change in subtidal rocky habitats'. *Global Change Biology*, vol 15, no 9, pp. 2153 2162.

Russell, L.K., Hepburn, C.D., Hurd, C.L. and Stewart, M.D. (2007). 'The expanding range of *Undaria pinnatifida* in southern New Zealand: Distribution, dispersal mechanisms and the invasion of wave-exposed environments'. *Biological Invasions*, vol 10, no 1, pp. 103–115.

Shears, N. and Bowen (2017). 'Half a century of coastal temperature records reveal complex warming trends in western boundary currents'. *Scientific Reports*, vol 7, p. 14527.

Smale, D.A., Burrows, M.T., Moore, P., O'Connor, N. and Hawkins, S.J. (2013). 'Threats and knowledge gaps for ecosystem services provided by kelp forests: A northeast Atlantic perspective'. *Ecology and Evolution*, vol 3, no 11, pp. 4016–4038.

Steneck, R. and Graham, M. (2002). 'Kelp forest ecosystems: Biodiversity, stability, resilience and future'. *Environmental Conservation*, vol 29, no 4, pp. 436–459.

Suárez R.J., Hepburn C.D., Hyndes G.A., McLeod R.J., Taylor R.B. and Hurd C.L. (2015). 'Contributions of an annual invasive kelp to native algal assemblages: Algal resource allocation and seasonal connectivity across ecotones'. *Phycologia*, vol 54, no 5, pp. 530–544.

Suárez-Jiménez, R., Hepburn, C.D., Hyndes, G.A., McLeod R.J., Taylor R.B. and Hurd, C.L. (2017). 'Importance of the invasive macroalga *Undaria pinnatifida* as trophic subsidy for a beach consumer'. *Marine Biology*, vol 164, no 5, p. 113.

Turley, C., Blackford, J., Widdicombe, S., Lowe, D., Nightingale, P.D. and Rees, A.P. (2006). 'Reviewing the impact of increased atmospheric CO_2 on oceanic pH and the marine ecosystem'. *Avoiding Dangerous Climate Change*, vol 8, pp. 65–70.

Turner, N.J., Berkes, F., Stephenson, J. and Dick, J. (2013). 'Blundering intruders: Extraneous impacts on two indigenous food systems'. *Human Ecology*, vol 41, no 4, pp. 563–574.

Vermeer, M. and Rahmstorf, S. (2009). 'Global sea level linked to global temperature'. *Proceedings of the National Academy of Sciences*, vol 106, no 51, pp. 21527–21532.

Watson, R.A., Cheung, W.W.L., Anticamara, J.A., Sumaila, R.U., Zeller, D. and Pauly, D. (2012). 'Global marine yield halved as fishing intensity redoubles'. *Fish and Fisheries*, vol 14, no 4, pp. 493–503.

Waycott, M., Duarte, C.M., Carruthers, T.J.B., Orth, R.J., Dennison, W.C., Olyarnik, S., Calladine, A., Fourqurean, J.W., Heck, Jr. K.L., Hugh, A.R., Kendrick G.R., Kenworthy, W.J., Short, F.T. and Williams, S.L. (2009). 'Accelerating loss of seagrasses across the globe threatens coastal ecosystems'. *Proceedings of the National Academy of Sciences*, vol 106, no 30, pp. 12377–12381.

Wernberg, T., Smale, D.A., Tuya, F., Thomsen, M.S., Langlois, T.J., de Bettignies, T., Bennett, S. and Rousseaux C.S. (2012). 'An extreme climatic event alters marine ecosystem structure in a global biodiversity hotspot'. *Nature Climate Change*, vol 3, no 1, pp. 78–82.

Win, R. (2011). 'The importance of macroalgae on rocky reefs: A critical aspect for fish and epifauna of the East Otago coastline'. MSc thesis, University of Otago, New Zealand.

Worm, B., Barbier, E.B., Beaumont, N., Duffy, J.E., Folke, C., Halpern, B.S., Jackson, J.B.C., Lotze, H.K., Micheli, F., Palumbi, S.R., Sala, E., Selkoe, K.A., Stachowicz, J.J. and Watson R. (2006). 'Impacts of biodiversity loss on ocean ecosystem services'. *Science*, vol 314, no 5800, pp. 787–790.

Part IV

Wealth from the oceans

10 Aquaculture

Andrew W.M. Pomeroy and Ana M.M. Sequeira

Introduction

Around the world, aquaculture is being viewed as an essential part of the solution to address the food security concerns associated with increased population. In 2013, for the first time, aquaculture provided more finfish for human consumption than capture fisheries (FAO, 2016). While the expansion of finfish aquaculture has been notable (5.9 per cent between 2001 and 2015 and as high as 10.8 per cent in the 1980s) (FAO, 2017), expansion in the production of molluscs, crustaceans and other aquatic animal species such as sea cucumbers has also occurred (FAO, 2016). Aquaculture also supplies commodities such as pearls and seaweed, and recently there has also been an increased effort to use aquaculture techniques to restore a range of ecosystems: notable examples are the re-population of seagrass meadows (e.g. Kenworthy *et al.*, 2007) and the cultivation of coral that is then transplanted onto degraded reefs (e.g. Young *et al.*, 2012).

Aquaculture production occurs at multiple scales, ranging from small family-owned operations to large multinational businesses. To address the challenge of food security, there is the need to continue to upscale aquaculture operations. However, upscaling aquaculture to provide the quantities required to feed the growing population poses a number of challenges. For example, pollution that originates from the oversupply of feed, as well as aquaculture faeces in particular, has also been shown to have impacts on benthic communities at various spatial scales (e.g. Law *et al.*, 2014) and on the physio-chemical composition of the water column (e.g. Tovar *et al.*, 2000). This means that aquaculture upscaling can lead to serious impacts on ecosystems, including: the alteration and destruction of habitats (e.g. Páez-Osuna, 2001), threats to aquatic biodiversity (e.g. Diana, 2009) and the spread of disease and genetic pollution (e.g. Bondad-Reantaso *et al.*, 2005; Stentiford *et al.*, 2012).

Moreover, as the aquaculture industry increases globally, and existing operations are intensified, the industry is faced with a number of challenges that range from the supply of adequate and sustainable feedstock to the impacts of the changing ocean climate. This chapter reviews the *extreme impacts* of intensive aquaculture on the marine environment, how *extreme marine conditions* associated with a changing climate may affect future aquaculture production and what *extreme economic, logistical and physical challenges* need to be overcome to expand aquaculture production into new regions.

A brief history of aquaculture

The practice of cultivating marine organisms has existed for thousands of years. For example, it has been suggested that as early as 6600 BC, the indigenous Gunditjmara people in Victoria, Australia, harvested eels through intensive manipulation of water courses in the landscape to create a complex system of 'stone-walled traps' (e.g. McNiven et al., 2012). Modern aquaculture, however, has long been credited to China, where aquaculture techniques have been documented at around 1126 BC (e.g. Nash, 2010). These systematic aquaculture methods principally focussed on the cultivation of the common carp in trapped lakes, and the widespread adoption of these methods is attributed to population pressures in the coastal zone. The techniques employed by the Chinese spread to Japan (e.g. Drews, 1951) and then to other parts of Asia (e.g. McLarney, 1984). Similar techniques appear also to have developed independently in ancient Rome (e.g. Balon, 1995), where fish and crustaceans were caught and then stored alive in tank-type structures until consumed. These freshwater techniques were refined and then expanded throughout Europe with the aid of treatises that provided details on construction and management techniques, as well as on the choice of species to farm, their diseases and their diet.

Mariculture, aquaculture based on seawater farming, dates back to at least the thirteenth century when mussels are first thought to have been cultured in France (e.g. Bardach et al., 1972). It has been speculated that, around the same time, systematic aquaculture systems were developed in Hawaii (e.g. Costa-Pierce, 1987). These aquaculture systems consisted of freshwater, brackish water and oceanic fish ponds constructed from rock and fill and used to cultivate a wide range of marine species. As aquaculture expanded, new marine species were trialled and by the late 1600s seaweed was being cultivated in Japan (e.g. Nash, 2010). However, it was not until the first half of the eighteenth century that fish farming in its modern form began when Stephan Ludwig Jacobi fertilised trout eggs and harvested fish that had been grown to maturity.

Aquaculture rapidly expanded in the nineteenth century as the hatchery process was refined and commercialised, and new techniques developed to expand the range of species that could be cultivated. In addition, public concerns about overexploitation of popular marine species and increased demand for fish, while wild catch fisheries have reduced, has encouraged aquaculture diversification. Consequently, of the species that had been cultured up to 2007, ~430 (97 per cent) were domesticated during the twentieth and twenty-first centuries, and 106 in the previous decade (Duarte et al., 2007).

Today, aquaculture operations occur as subsistence farms to large multi-national companies all over the world, and a range of technical approaches are used. However, while further expansion of aquaculture seems inevitable, aquaculture is at risk from its own development as well as a changing climate.

Extreme impacts

As the importance of aquaculture grows and production increases, the environmental impacts are being closely scrutinised. This is in part due to the increased prevalence of aquaculture operations but also due to a few, but well publicised, cases of substantial environmental degradation. As aquaculture production intensifies and expands globally, the impacts of the industry on the environment will continue to increase. Considerable research has been, and continues to be, undertaken to understand these impacts and to identify novel solutions to address them.

Environmental impact

Aquaculture will inevitably have an impact on the receiving environment. Thus, there is a need to manage this impact as aquaculture production expands and production density increases. Marine aquaculture typically consists of three production methods: cage or pen culture for finfish; culture of bivalve molluscs and seaweed near the shoreline on long-lines or in basket-type structures; and ranching activity. Each of these structures can alter hydrodynamic processes (i.e. waves and currents), which in turn can affect the capacity of the physical processes that are relied upon to deliver water of suitable quality and to disperse waste produced by the marine species being cultivated (e.g. Lin et al., 2016). Consequently, waste deposition under finfish and mollusc aquaculture, in particular, can be substantial.

Aquaculture contributes organic and inorganic compounds into the water, mostly in the form of animal faeces and feed waste. As aquaculture density increases, the percentage of dissolved oxygen in the water can be reduced (e.g. Hargrave et al., 1993). Under certain conditions, this reduction can be rapid and may cause mortality. Furthermore, the ratio of nitrogen to phosphorous can also be affected. In addition to changes in the water quality, the deposition of waste on the seabed changes the sediment chemistry and the composition of the benthic communities (e.g. Karakassis et al., 1999), sometimes leading to the proliferation of tolerant and opportunistic species. A number of studies have demonstrated that microbiological metabolism in sediments is stimulated by organic wastes from cages and can be as high as 10 times greater than nearby reference sites (e.g. Holmer and Kristensen, 1992), and benthos activity in the vicinity of a cage has been shown to be 4–6 times higher than reference sites. However, while activity may increase, the overall state of the benthos community may still be negatively affected or, under some circumstances, it may disappear completely (e.g. Pergent-Martini et al., 2006).

In contrast to finfish aquaculture, bivalve aquaculture has been shown to filter substantial quantities of water, thus playing an important role in the denitrification of waters (e.g. Kaspar *et al.*, 1985). However, like finfish, bivalve aquaculture has also been shown to affect the receiving environment through increased demand for dissolved oxygen and elevated levels of organic matter deposition. Heavy sedimentation of faeces has been recorded beneath mussel farms, with sedimentation rates up to three times higher than reference sites (e.g. Chamberlain *et al.*, 2001).

A number of studies have suggested that the environmental impact of the waste from aquaculture is restricted to the immediate vicinity of the operations and decreases substantially with distance from the production sites. However, as production density increases to extreme levels and more regions are developed, the cumulative impact can be expected to be far greater than that inflicted by a single farm operation. Moreover, oceanographic conditions and currents regimes play an important role in the spread of farm residuals. Such density effects remain poorly quantified.

Production connectivity

Production density, both in terms of site distribution and intensiveness, can increase the connectivity between aquaculture operations. As aquaculture production expands to extreme levels, increase connectivity results in individual farms no longer operating in isolation; the environmental impacts from one aquaculture operation will also affect the environment of the neighbouring operation. This connectivity reduces the resilience of each operation and increases the risk and impact of disease.

Stress on fish that results from high density and intensive cultivation is often sufficient to allow opportunistic pathogens to infect cultured fish stock. For many aquatic species, this stress is due to temperature fluctuations, changes in salinity and pH, low dissolved oxygen and algal blooms (e.g. Snieszko, 1974; Harvell *et al.*, 1999). Thus, increased water temperatures as well as changes in the water composition can be expected to alter the incidence of diseases. Such diseases can be caused either by opportunistic pathogens that are widespread and can infect the host when its resistance is reduced, or by obligate pathogens that occur in the presence of a suitable host and spread through farm connectivity as well as from foreign locations through increased international trade (e.g. Bernoth *et al.*, 2008). Once cultured stock is infected, disease can proliferate quickly within and between farm operations that are oceanographically connected (e.g. farms closely connected by currents) (e.g. Viljugrein *et al.*, 2009; Salama and Murray, 2011).

Several vectors for the transfer of diseases between cultured and native species have also been identified (e.g. Johansen *et al.*, 2011). While escaped animals receive the highest media attention due to their potential to cause genetic pollution (e.g. Crozier, 1993), intermediate hosts can also be important vectors (e.g. Arechavala-Lopez *et al.*, 2013) – for example, wild fish that move in and out of aquaculture farms but may themselves not be affected. Such wild fish can rapidly spread disease over large spatial areas. As aquaculture production increases, there is a need for a joint strategy to be developed between farms in order to coordinate fallowing (the aquaculture equivalent to rotating crops) of regions to minimise disease transfer as well as enable the sedimentary environment to recover (e.g. Pereira *et al.*, 2004).

Adaptive management and spatial planning approaches

Rapid growth in aquaculture production has been observed in countries where methodologies for zoning and licensing are in place. However, the spatial planning process that underpins this zoning and licensing is often not straightforward. As the demand for space increases, particularly in coastal areas, intensification of aquaculture competes with a range of other industries (e.g. fishing, tourism and recreation, renewable energies). In addition to competition for space, there is greater scrutiny on all users of the oceans. Consequently, there are increasing environmental, economic and policy compliance related issues that also need to be considered (e.g. Galparsoro *et al.*, 2018) and it is well recognised that the spatial planning process is improved when embedded in an ecosystem approach (e.g. Halpern *et al.*, 2010; Katsanevakis *et al.*, 2011).

The implementation of an Ecosystem Approach to Aquaculture (EAA) has been promoted by the Food and Agriculture Organization (FAO), which suggested four governance principles for the sustainable development of aquaculture: accountability; effectiveness and efficiency of governments; equity; and predictability (e.g. Soto *et al.*, 2008; Hishamunda *et al.*, 2014). A number of methodologies have been developed to try to address the different issues associated with aquaculture development. For example, Integrated Coastal Zone Management (ICZM) seeks to administer multiple sectors and policies within coastal zones (e.g. Katsanevakis *et al.*, 2011); and guidelines on how to allocate zones for aquaculture have been suggested by the General Fisheries Commission for the Mediterranean (GFCM, 2012). However, implementation of such planning methodologies requires a thorough account of the potential impacts associated with levels of production and density in order to determine their cumulative effects on adjacent areas (e.g. Marine Strategy Framework Directive: Directive 2008/56/EC). Due to constraints associated with addressing these issues, lack of space is a common factor hindering aquaculture development, especially for finfish (e.g. Hofherr *et al.*, 2015). To address this issue, a range of measures are being explored, including the use of buffering zones or the integration of multi-trophic aquaculture (i.e. the joint culture of finfish, seaweed or shellfish) (e.g. Ridler *et al.*, 2007).

The biomass imbalance (more feed to produce less fish)

The global demand for food protein has followed the exponential increase in human population (e.g. Godfray *et al.*, 2010). Aquaculture is expected to provide a considerable proportion of this protein demand as wild fisheries catches decrease globally (FAO, 2016). However, this reliance on aquaculture for feeding the world's increasing population means that for many types of aquaculture, based on the feed presently used, more natural resources will be required to produce that feed (as stated by Pahlow *et al.*, 2015). An important threshold for fish production is the feed conversion ratio, which corresponds to the ratio between the amount of protein used in the feed and the protein in the cultured fish. For example, in the well-developed salmon aquaculture sector, the ratio has currently been optimised to be close to one (e.g. Fry *et al.*, 2018); considerable focus is also being placed on other high-value fish such as tuna. While obtaining feed conversion ratios around one are highly desirable, one of the greatest challenges facing aquaculture sustainability is the composition of feed. Production of aquaculture feed currently relies on wild fish protein and oil and represents, therefore, a great risk of overexploiting fish species that are used as feed. Thus, as aquaculture expands around the world, promoting the cultivation of low trophic level species for protein production as well as finding alternative sources of protein and oil for fish feed is necessary, and this is being investigated (e.g. use of insects) (Barroso *et al.*, 2014).

Extreme marine conditions

Marine aquaculture development relies on an ocean environment that is suitable for predictable, efficient and sustainable production. However, as the global climate changes, variability in temperature and rainfall patterns, combined with increased uptake of carbon dioxide by the oceans, changes in currents and wave regimes and changes in biogeochemical properties of the oceans are expected. These changes are likely to be enhanced in shallow coastal environments, particularly where circulation is limited. Furthermore, greater energy in the ocean due to increased and higher intensity storms may have implications for aquaculture infrastructure as well as the safety and security of some aquaculture operations. In this section, the impacts on marine aquaculture of extreme marine conditions due to changes in the global climate are discussed.

Changing temperatures

Increased sea surface temperature (SST), combined with seasonal variability, may have both positive and negative implications for aquaculture. SST is the water temperature close to the sea surface and typically represents the temperature of the upper mixed layer, within which most forms of aquaculture take place. Climate models, as well as direct measurements, demonstrate that SST is increasing in many parts of the world (IPCC, 2014). The most recent analysis indicates that the global average SST has warmed by +0.11°C per decade at least since 1971 and continues to accelerate (IPCC, 2014). While considerable effort is being made to address anthropogenic contributors to climate change, such as greenhouse gas emissions, the thermal inertia of the oceans is expected to result in increases in SST for several decades yet (Meehl *et al.*, 2005; IPCC, 2013).

For finfish aquaculture, moderate increases in SST are most likely to have implications for the collection of wild juvenile stock that is then grown-out in aquaculture farms. This is due to the impacts on rate and duration of spawning occurrence as well as survival (e.g. Wainwright, 1982). Juvenile stock produced in hatcheries is less likely to be affected as these operations are typically temperature controlled. Once transferred to sea-cages for grow-out, moderate increases in SST have been suggested to raise the metabolic rate for some finfish species, which may lead to higher growth rates and increase food conversion efficiency (e.g. Beamish, 1970; Handeland *et al.*, 2008). However, at sustained high SST or for short but extreme exposure, growth may be inhibited (e.g. Hokanson *et al.*, 1977) or mortality may occur.

Similar impacts are expected from warmer SST for other aquatic species. For example, geographic differences in SST have been related to variability in the production of sea cucumbers (e.g. Purcell and Simutoga, 2008). Variations in SST have also been shown to contribute to mortality of oysters (e.g. Morizane *et al.*, 2001; O'Connor and Lawler, 2004) and to affect the quality of the pearl produced, which depends on the thickness and deposition rate of the nacre. This deposition rate has been shown to be correlated with increases in water temperature and it is generally accepted that higher quality nacre with superior lustre is deposited in cool waters (e.g. Gervis and Sims, 1992).

The largest impact of increased SST will be on the cultivation of seaweed. Many of the cultured seaweed species, often produced in the reef lagoons of small island developing nations, are already near the upper limit of their temperature tolerance (e.g. Pickering, 2006) and their capacity to adapt remains unclear (e.g. Harley *et al.*, 2012). At high water temperatures, many seaweed species experience stress that inhibits growth (e.g. Buschmann *et al.*, 2004). While there is limited scope for adaptation of seaweed farming at the regional level by shifting production to higher latitudes or deeper waters, further research is required to identify alternative, commercially attractive species suitable for warmer water production. Similarly, coral used for ornamental trade as well as reef restoration projects is also typically cultured in shallow, sheltered coastal areas. At high temperatures, the major response of corals is to bleach by expelling their zooxanthellae (e.g. Glynn, 1993; Brown, 1997), which they need for energy production. If the bleaching is protracted, mortality occurs. This will impact the effectiveness of coral rearing projects as well as the survivability of coral that is transplanted onto degraded reef areas.

Water properties

As the global climate changes, the properties of coastal waters will also change. Of particular relevance for aquaculture production are changes in salinity and dissolved oxygen content, increases in turbidity and a decrease in pH (ocean acidity). In coastal waters, changes in temperature and rainfall affect the salinity of the water and, perhaps more importantly for aquaculture, can induce stratification. When coastal waters become stratified the level of dissolved oxygen is reduced, particularly in locations with poor circulation, and can reach anoxic conditions and cause mortality. Dissolved oxygen can also be rapidly depleted due to algal blooms, which can be associated with eutrophication (excess nutrients in the water) (e.g. Hallegraeff, 1993).

Waves and currents, as well as changes in rainfall patterns, can increase water turbidity. In many regions of the world, storminess is expected to escalate, which may increase sediment mobilisation and transport as well as land runoff. Sustained turbidity can lead to mass mortality of molluscs when suspended solids in turbid water exceeds the energy of the molluscs to filter those solids (e.g. Gervis and Sims, 1992). The implications of elevated turbidity for finfish are not well understood but are likely to be associated with changes in dissolved oxygen due to particle resuspension (e.g. Poizot *et al.*, 2016) rather than the suspension of the sediment itself.

As carbon dioxide concentrations increase in the atmosphere and are absorbed into ocean waters, the pH of coastal waters, where aquaculture occurs, is predicted to decrease (e.g. Doney *et al.*, 2009). This reduction in pH is expected to reduce calcification rates, affecting marine species that form shells or skeletons from calcium carbonate such as corals and molluscs (e.g. Hoegh-Guldberg *et al.*, 2007; Fabry *et al.*, 2008). For molluscs, this may result in an increase in the percentage of stock with thin and abnormal shells (e.g. Kurihara *et al.*, 2007), which may increase the risk of pathogens as well as mechanical damage (e.g. Dove and Sammut, 2007). For pearls, in particular, it is also likely that production quality will be reduced because the nacre of pearl oysters is mostly composed of the more soluble aragonite (e.g. Jacob *et al.*, 2008). However, the extent of this impact remains unclear due to the development of round pearls within oyster tissue that insulates them from the surrounding waters (e.g. Welladsen *et al.*, 2010). Similar impacts due to decreased pH are also expected for corals (Hoegh-Guldberg *et al.*, 2007), which may reduce the success of cultivation and restoration projects. While changes in pH have been shown to affect fish behaviour (e.g. Nilsson *et al.*, 2012), it remains unclear how this may affect the productivity of finfish aquaculture. It is possible that, for a change in pH of 0.1–0.3 units, the impact may be low for hatchery-based marine fish aquaculture that grow-out fish in sea cages (e.g. Bell *et al.*, 2011).

Increased ocean energy and sea level

Most aquaculture production and its associated infrastructure is located in coastal areas, which are expected to be subject to increased sea levels and extreme events (storminess) (IPCC, 2013). For coastal embayments, this may lead to increased circulation and changes in streamflow as well as changes in water properties (e.g. Najjar *et al.*, 2010). Similarly, increased water depths over a coastal reef will enable larger waves to propagate over the reef, which will increase currents and reduce the residence time of sea water in enclosed lagoons. These changes will be contingent on the structural complexity of the reef, which dissipates waves and attenuates currents in these environments (e.g. Harris *et al.*, 2018). In some cases, increased water depths may mitigate some of the impacts of increased SST (e.g. Lowe *et al.*, 2016).

For most forms of aquaculture, increased ocean energy and higher sea levels will predominantly affect farm infrastructure. Therefore, there will be a need to evaluate the forces imposed on such infrastructures, their elevation above mean sea level and the technologies used for construction and maintenance at the production site. The greatest impact will be on seaweed farming, which is typically undertaken in shallow, protected subtidal areas or in reef lagoons and is particularly affected by water motion (e.g. Lin *et al.*, 2016). Increased water depths may lead to some locations becoming unsuitable for seaweed farming. In contrast, where the depth remains suitable, moderate increases in local currents may improve water quality through increased nitrogen and suspended organic particles and assist in offsetting SST increases, which may benefit seaweed and mollusc production (e.g. Pouvreau *et al.*, 2006). For sea cucumber production, however, such impacts can cause mortality or product loss, as these species are highly sensitive to the impact of increased storminess (e.g. Purcell and Simutoga, 2008).

Extreme economic, logistical and physical challenges to expansion

As coastal aquaculture production is intensified, spatial conflicts between aquaculture production and other uses of the coastal zone is increasing. For example, there is often considerable controversy between proposed aquaculture projects and hotel owners who believe that fish farming will result in an increased occurrence of sharks in the area, affect the water quality and have a visual impact. To address these conflicts, and to provide greater operational space, aquaculture operators are increasingly looking to more exposed locations (e.g. Skladany *et al.*, 2007). While mollusc farming (e.g. mussels and pearls) is already taking place in offshore waters in a number of countries, most finfish farms presently operating offshore are still in the research and development phase. This is because the species that are suitable for offshore environments are still being identified and differences in stocking densities, feeding behaviours, feed formulations and food conversion ratios are being investigated. For example, while both salmon and tuna are grown in cooler temperatures (e.g. in the fjords in Norway, Tasmania and South Australia), the culture of barramundi in tropical waters is also becoming common (e.g. north of Australia). With the need to expand aquaculture production to secure food for the future, production of temperate water species, such as yellow tail kingfish (*Seriola lalandi*), is also now expanding (e.g. Fernandes and Tanner, 2008; Blanco Garcia *et al.*, 2015). However, the suitability of many different aquaculture products still needs to be assessed. For example, most seaweed culture methods are not designed for open seas (North, 1987) and will require modification (Buck and Buchholz, 2005).

For offshore farms to be economically viable, it is likely that they will need to be much larger than the currently existing nearshore farms. Consequently, more waste will be generated in each farm, although it remains unclear if the properties of this waste will differ from that produced in nearshore operations (e.g. Reid *et al.*, 2009). While faster currents, deeper water and lower nutrient baselines may reduce the local impacts from offshore operations, the overall capacity of the offshore ecosystem to assimilate production waste remains unclear, particularly in deeper waters where assimilation is expected to be reduced (e.g. Lin and Bailey-Brock, 2008). Furthermore, there is limited knowledge about linkages between offshore and nearshore systems and how these processes may affect biogeochemical cycling as well as ecology at a larger scale. Moreover, the offshore environment is increasingly being exploited for other uses, including maritime transport, energy and tourism, highlighting the need for careful planning taking into account the increasing demand for space (e.g. Douvere, 2008).

Expanding aquaculture into deeper water also has technological challenges. Setting aside the issues associated with competition for space with, for example, marine protected areas, there is also the need to find favourable 'operational characteristics' (Gentry *et al.*, 2017b, p. 735). Expansion into deeper water also requires new planning (e.g. Gentry *et al.*, 2017b) and technological solutions to providing aquaculture infrastructure (such as anchoring systems) that can resist the generally harsher offshore environment (e.g. Stevens *et al.*, 2008) as well as be more autonomous and require lower maintenance. For example, strong offshore currents and large waves may impose high forces on both the cultured species and the infrastructure (e.g. Buck and Buchholz, 2005). While suitable technology may exist in other fields (such as the offshore oil and gas industry), translating this technology and reducing the costs associated with their application will be essential; the cost structures between these industries are very different. Thus, it remains to be seen if offshore aquaculture will be economically viable and whether the increase in production scale will be sufficient to offset increases in the costs associated with offshore aquaculture (particularly for carnivorous finfish) such as feed, construction, maintenance and logistical costs.

The future

It is clear that as aquaculture production as well as marine conditions become more extreme, identification of sites and species suitable for aquaculture production, as well as new methods to address its impacts, will need to be developed. One such approach that has received considerable attention is the use of polyculture methods such as integrated multi-trophic aquaculture, which have been shown to mitigate some of the impacts while producing useful by-products (e.g. Chopin *et al.*, 2001). These methods rely on combining extractive species with other forms of aquaculture (e.g. finfish with mussels and seaweed). While this concept is attractive, it remains unclear how polyculture can contribute to the overall economic performance, both for nearshore and offshore systems. It has been proposed that in order to expand the use of polyculture there is a need to identify and quantify the environmental costs of cultivating feed species, and to find ways to economically quantify the positive effects of integration with extractive species. In addition, it is still necessary to know what values of ecosystem goods and services are being generated from natural ecosystems, and how aquaculture wastes affect them. Such information is scarce, particularly for offshore environments. Finally, with increased interest in the development of offshore aquaculture, the economics of integrating aquaculture with other offshore operations, such as wind farms, is being investigated (e.g. Buck *et al.*, 2008). Such integration will assist in reducing the costs of offshore farm development and operation as well as limiting the overall footprint of offshore development more generally.

References

Arechavala-Lopez, P., Sanchez-Jerez, P., Bayle-Sempere, J.T., Uglem, I. and Mladineo, I. (2013) 'Reared fish, farmed escapees and wild fish stocks – A triangle of pathogen transmission of concern to Mediterranean aquaculture management'. *Aquaculture Environment Interactions*, vol 3, no 2, pp. 153–161. doi:10.3354/aei 00060.

Balon, E.K. (1995) 'Origin and domestication of the wild carp, *Cyprinus carpio*: From Roman gourmets to the swimming flowers'. *Aquaculture*, vol 129, no 1–4, pp. 3–48. doi:10.1016/0044-8486(94)00227-F.

Bardach, J.E., Ryther, J.H. and McLarney, W.O. (1972) *Aquaculture: The Farming and Husbandry of Freshwater and Marine Organisms*. New York: John Wiley & Sons.

Barroso, F.G., de Haro, C., Sánchez-Muros, M.J., Venegas, E., Martínez-Sánchez, A. and Pérez-Bañón, C. (2014) 'The potential of various insect species for use as food for fish'. *Aquaculture*, vol 422, pp. 193–201. doi:10.1016/j.aquaculture.2013.12.024.

Beamish, F.W.H. (1970) 'Influence of temperature and salinity acclimation on temperature preferenda of the euryhaline fish *Tilapia nilotica*'. *Journal of the Fisheries Research Board of Canada*, vol 27, no 7, pp. 1209–1214, doi:10.1139/f70-143.

Bell, J.D., Johnson, J.E. and Hobday, A.J. (eds.) (2011) 'Vulnerability of tropical pacific fisheries and aquaculture to climate change'. Secretariat of the Pacific Community, Noumea, New Caledonia.

Bernoth, E.M, Chavez, C, Chinabut, S and Mohan, C.V. (2008) 'International trade in aquatic animals – A risk to aquatic animal health status', in M.G. Bondad-Reantaso, C.V. Mohan, M. Crumlish and R.P Subasinghe (eds.), 'Diseases in Asian aquaculture VI', pp. 53–70. Fish Health Section, Asian Fisheries Society, Manila.

Blanco Garcia, A., Partridge, G.J., Flik, G., Roques, J.A.C. and Abbink, W. (2015) 'Ambient salinity and osmoregulation, energy metabolism and growth in juvenile yellowtail kingfish (*Seriola lalandi* Valenciennes 1833) in a recirculating aquaculture system'. *Aquaculture Research*, vol 46, no 11, pp. 2789–2797. doi:10.1111/are. 12433.

Bondad-Reantaso, M.G., Subasinghe, R.P., Arthur, J.R., Ogawa, K., Chinabut, S., Adlard, R., Tan, Z. and Shariff, M. (2005) 'Disease and health management in Asian aquaculture'. *Veterinary Parasitology*, vol 132, no 3–4 pp. 249–272. doi:10.1016/j.vetpar.2005.07.005.

Brown, B.E. (1997) 'Coral bleaching: Causes and consequences'. *Coral Reefs*, vol 16, no 1, S129–S138. doi:10.1007/s003380050249.

Buck, B.H. and Buchholz, C.M. (2005) 'Response of offshore cultivated *Laminaria saccharina* to hydrodynamic forcing in the North Sea'. *Aquaculture*, vol 250, no 3–4, pp. 674–691. doi:10.1016/j.aquaculture.2005.04.062.

Buck, B.H., Krause, G., Michler-Cieluch, T., Brenner, M., Buchholz, C.M., Busch, J.A., Fisch, R., Geisen, M. and Zielinski, O. (2008) 'Meeting the quest for spatial efficiency: Progress and prospects of extensive aquaculture within offshore wind farms'. *Helgoland Marine Research*, vol 62 no 3, pp. 269–281. doi:10.1007/s10152-008-0115-x.

Buschmann, A.H., Vásquez, J.A., Osorio, P., Reyes, E., Filún, L., Hernández-González, M.C. and Vega, A. (2004) 'The effect of water movement, temperature and salinity on abundance and reproductive patterns of *Macrocystis* spp. (Phaeophyta) at different latitudes in Chile'. *Marine Biology*, vol 145, no 5, pp. 849–862. doi:10.1007/s00227-004-1393-8.

Chamberlain, J., Fernandes, T.F., Read, P., Nickell, T.D. and Davies, I.M. (2001) 'Impacts of biodeposits from suspended mussel (*Mytilus edulis* L.) culture on the surrounding surficial sediments'. *ICES Journal of Marine Science*, vol 58, no 2, pp. 411–416. doi:10.1006/jmsc.2000.1037.

Chopin, T., Buschmann, A.H., Halling, C., Troell, M., Kautsky, N., Neori, A., Kraemer, G.P., Zertuche-González, J.A., Yarish, C. and Neefus, C. (2001) 'Integrating seaweeds into marine aquaculture systems: A key toward sustainability'. *Journal of Phycology*, vol 37, no 6, pp. 975–986. doi:10.1046/j.1529-8817.2001.01137.

Costa-Pierce, B.A. (1987) 'Aquaculture in ancient Hawaii'. *BioScience*, vol 37, no 5, pp. 320–331. doi:10.2307/1310688.

Crozier, W.W. (1993) 'Evidence of genetic interaction between escaped farmed salmon and wild Atlantic salmon (*Salmo salar* L.) in a Northern Irish river'. *Aquaculture*, vol 113, no 1–2, pp. 19–29. doi:10.1016/0044-8486(93)90337-X.

Diana, J.S. (2009) 'Aquaculture production and biodiversity conservation'. *BioScience*, vol 59, no 1, pp. 27–38. doi:10.1525/bio.2009.59.1.7.

Doney, S.C., Fabry, V.J., Feely, R.A. and Kleypas, J.A. (2009) 'Ocean acidification: The other CO_2 problem'. *Annual Review of Marine Science*, vol 1, no 1, pp. 169–192. doi:10.1146/annurev.marine.010908.163834.

Douvere, F. (2008) 'The importance of marine spatial planning in advancing ecosystem-based sea use management'. *Marine Policy*, vol 32, no 5, pp. 762–771. doi:10.1016/j.marpol.2008.03.02.

Dove, M.C. and Sammut, J. (2007) 'Impacts of estuarine acidification on survival and growth of Sydney rock oysters *Saccostrea glomerata* (Gould 1850)', *Journal of Shellfish Research*, vol 26, no 2, pp. 519–527. doi:10.2983/0730-8000(2007)26[519:IOEAOS] 2.0.CO;2.

Drews, R.A. (1951) 'The cultivation of food fish in China and Japan: A study disclosing contrasting national patterns for rearing fish consistent with the differing cultural histories of China and Japan'. PhD thesis, University of Michigan.

Duarte, C.M., Marbá, N. and Holmer, M. (2007) 'Rapid domestication of marine species'. *Science*, vol 316, no 5823, pp. 382–383. doi:10.1126/science.1138042.

Fabry, V.J., Seibel, B.A., Feely, R.A. and Orr, J.C. (2008) 'Impacts of ocean acidification on marine fauna and ecosystem processes'. *ICES Journal of Marine Science*, vol 65, no 3, pp. 414–432. doi:10.1093/icesjms/fsn048.

FAO (2016) 'The state of world fisheries and aquaculture 2016. Contributing to food security and nutrition for all'. Food and Agriculture Organization of the United Nations, Rome.

FAO (2017) 'FAO yearbook. Fishery and aquaculture statistics. FAO annuaire. Statistiques des pêches et de l'aquaculture. FAO anuario. Estadísticas de pesca y acuicultura. 2015'. Food and Agriculture Organization of the United Nations, Rome.

Fernandes, M. and Tanner, J. (2008) 'Modelling of nitrogen loads from the farming of yellowtail kingfish *Seriola lalandi* (Valenciennes, 1833)'. *Aquaculture Research*, vol 39, no 12, pp. 1328–1338. doi:10.1111/j.1365-2109.2008.02001.x.

Fry, J.P., Mailloux, N.A., Love, D.C., Milli, M.C. and Cao, L. (2018) 'Feed conversion efficiency in aquaculture: Do we measure it correctly?'. *Environmental Research Letters*, vol. 13, no. 2, p. 024017. doi:10.1088/1748-9326/aaa273.

Galparsoro, I., Murillas, A., Pinarbasi, K., Borja, A.Á., O'Hagan, M., MacMahon, E., Sequeria, A.M., Gangnery, A., Corner, R., Ferreira, J., Ferreira, R., Gimpel, A., Boyd, A., Icely, J., Bergh, Ø., Donohue, C., Lui, H., Billing, S., Garmendia, J.M., Lagos, L. and Arantzamendi, L. (2018) 'Synthesis of the lessons learned from the development and testing of innovative tools to support ecosystem-based spatial planning to aquaculture'. Deliverable 5.1, AquaSpace: Ecosystem approach to making space for aquaculture, EU Horizon 2020 project grant no 633476.

Gentry, R.R., Froehlich, H.E., Grimm, D., Kareiva, P., Parke, M., Rust, M., Gaines, S.D. and Halpern, B.S. (2017a) 'Mapping the global potential for marine aquaculture'. *Nature Ecology & Evolution*, vol 1, no 9, pp. 1317–1324. doi:10.1038/s41559-017-0257-9.

Gentry, R.R., Lester, S.E., Kappel, C.V., White, C., Bell, T.W., Stevens, J. and Gaines, S.D. (2017b) 'Offshore aquaculture: Spatial planning principles for sustainable development'. *Ecology and Evolution*, vol 7, no 2, pp. 733–743. doi:10.1002/ece3.2637.

Gervis, M.H. and Sims, N.A. (1992) 'The biology and culture of pearl oysters (Bivalvia: Pteriidae)'. *ICLARM Studies and Reviews*, vol 21, pp.1–49.

GFCM (2012) 'Resolution GFCM/36/2012/1 on guidelines on Allocated Zones for Aquaculture (AZA)'. General Fisheries Commission for the Mediterranean, 36th Session, Marrakech, Morocco, 14–19 May 2012. Available at: www.fao.org/gfcm/activities/aquaculture/projects/shocmed/en/.

Glynn, P.W. (1993) 'Coral reef bleaching: Ecological perspectives'. *Coral Reefs*, vol 12, no 1, pp. 1–17. doi:10.1007/BF00303779.

Godfray, H.C.J., Beddington, J.R., Crute, I.R., Haddad, L., Lawrence, D., Muir, J.F., Pretty, J., Robinson, S., Thomas, S.M. and Toulmin, C. (2010) 'Food security: The challenge of feeding 9 billion people'. *Science*, vol 327, no 5967, pp. 812–818. doi:10.1126/science.1185383.

Hallegraeff, G.M. (1993) 'A review of harmful algal blooms and their apparent global increase'. *Phycologia*, vol. 32, no. 2, pp. 79–99. doi:10.2216/i0031-8884-32-2-79.1.

Halpern, B.S., Lester, S.E. and McLeod, K.L. (2010) 'Placing marine protected areas onto the ecosystem-based management seascape'. *Proceedings of the National Academy of Sciences of the United States of America*, vol 107, no 43, pp. 18312–18317. doi:10.1073/pnas.0908503107.

Handeland, S.O., Imsland, A.K. and Stefansson, S.O. (2008) 'The effect of temperature and fish size on growth, feed intake, food conversion efficiency and stomach evacuation rate of Atlantic salmon post-smolts'. *Aquaculture*, vol 283, no 1–4, pp. 36–42. doi:10.1016/j.aquaculture.2008.06.042.

Hargrave, B.T., Duplisea, D.E., Pfeiffer, E. and Wildish, D.J. (1993) 'Seasonal changes in benthic fluxes of dissolved oxygen and ammonium associated with marine cultured Atlantic salmon'. *Marine Ecology Progress Series*, vol 96, no 3, pp. 249–257. Available at: www.jstor.org/stable/24833553.

Harley, C.D., Anderson, K.M., Demes, K.W., Jorve, J.P., Kordas, R.L., Coyle, T.A. and Graham, M.H. (2012) 'Effects of climate change on global seaweed communities'. *Journal of Phycology*, vol 48, no 5, pp. 1064–1078. doi:10.1111/j.1529-8817.2012.01224.x.

Harris, D.L., Rovere, A., Casella, E., Power, H., Canavesio, R., Collin, A., Pomeroy, A., Webster, J.M. and Parravicini, V. (2018) 'Coral reef structural complexity provides important coastal protection from waves under rising sea levels'. *Science Advances*, vol 4, no 2. doi:10.1126/sciadv.aao4350.

Harvell, C.D., Kim, K., Burkholder, J.M., Colwell, R.R., Epstein, P.R., Grimes, D.J., Hofmann, E.E., Lipp, E.K., Osterhaus, A.D.M.E, Overstreet, R.M., Porter, J.W., Smith, G.W. and Vasta, G.R. (1999) 'Emerging marine diseases – Climate links and anthropogenic factors'. *Science*, vol 285, no 5433, pp. 1505–1510. doi:10.1126/science.285.5433.1505.

Hishamunda, N., Ridler, N. and Martone, E. (2014) 'Policy and governance in aquaculture: Lessons learned and way forward'. FAO Fisheries and Aquaculture technical paper 577, Food and Agriculture Organization of the United Nations, Rome.

Hoegh-Guldberg, O., Mumby, P.J., Hooten, A.J., Steneck, R.S., Greenfield, P., Gomez, E., Harvell, C.D., Sale, P.F., Edwards, A.J., Caldeira, K., Knowlton, N., Eakin, C.M., Iglesias-Prieto, R., Muthiga, N., Bradbury, R.H., Dubi, A. and Hatziolos, M.E. (2007) 'Coral reefs under rapid climate change and ocean acidification', *Science*, vol 318, no 5857, pp. 1737–1742. doi:10.1126/science.1152509.

Hofherr, J., Natale, F. and Trujillo, P. (2015) 'Is lack of space a limiting factor for the development of aquaculture in EU coastal areas?'. *Ocean & Coastal Management*, vol 116, pp. 27–36. doi:10.1016/j.ocecoaman.2015.06.010.

Hokanson, K.E.F., Kleiner, C.F. and Thorslund, T.W. (1977) 'Effects of constant temperatures and diel temperature fluctuations on specific growth and mortality rates and yield of juvenile rainbow trout, *Salmo gairdneri*'. *Journal of the Fisheries Research Board of Canada*, vol 34, no 5, pp. 639–648. doi:10.1139/f77-100.

Holmer, M. and Kristensen, E. (1992) 'Impact of marine fish cage farming on metabolism and sulfate reduction of underlying sediments'. *Marine Ecology Progress Series*, vol 80, no 2–3, pp. 191–201. Available at: www.jstor.org/stable/24826606.

IPCC (2013) *Climate Change 2013: The Physical Science Basis. Contribution of Working Group I to the Fifth Assessment Report of the Intergovernmental Panel on Climate Change*

[T.F Stocker, D. Qin, G.-K. Plattner, M. Tignor, S.K. Allen, J. Boschung, A. Nauels, Y. Xia, V. Bex and P.M. Midgley (eds.)]. Cambridge and New York: Cambridge University Press.

IPCC (2014) 'Climate change 2014: Synthesis report. Contribution of working groups I, II and III to the fifth assessment report of the Intergovernmental Panel on Climate Change' [Core writing team, R.K. Pachauri and L.A. Meyer (eds.)]. Intergovernmental Panel on Climate Change, Geneva.

Jacob, D.E., Soldati, A.L., Wirth, R., Huth, J., Wehrmeister, U. and Hofmeister, W. (2008) 'Nanostructure, composition and mechanisms of bivalve shell growth'. *Geochimica et Cosmochimica Acta*, vol 72, no 22, pp. 5401–5415. doi:10.1016/j.gca.2008.08.019.

Johansen, L.H., Jensen, I., Mikkelsen, H., Bjørn, P.A., Jansen, P.A. and Bergh, Ø. (2011) 'Disease interaction and pathogens exchange between wild and farmed fish populations with special reference to Norway'. *Aquaculture*, vol 315, no 3–4, pp. 167–186. doi:10.1016/j.aquaculture.2011.02.014.

Karakassis, I., Hatziyanni, E., Tsapakis, M. and Plaiti, W. (1999) 'Benthic recovery following cessation of fish farming: A series of successes and catastrophes'. *Marine Ecology Progress Series*, vol 184, pp. 205–218. Available at: www.jstor.org/stable/24853244.

Kaspar, H.F., Gillespie, P.A., Boyer, I.C. and MacKenzie, A.L. (1985) 'Effects of mussel aquaculture on the nitrogen cycle and benthic communities in Kenepuru Sound, Marlborough Sounds, New Zealand'. *Marine Biology*, vol 85, no 2, pp. 127–136. doi:10.1007/BF00397431.

Katsanevakis, S., Stelzenmüller, V., South, A., Sørensen, T.K., Jones, P.J.S., Kerr, S., Badalamenti, F., Anagnostou, C., Breen, P., Chust, G., D'Anna, G., Duijn, M., Filatova, T., Fiorentino, F., Hulsman, H., Johnson, K., Karageorgis, A.P., Kröncke, I., Mirto, S., Pipitone, C., Portelli, S., Qiu, W., Reiss, H., Sakellariou, D., Salomidi, M., van Hoof, L., Vassilopoulou, V., Vega Fernández, T., Vöge, S., Weber, A., Zenetos, A. and ter Hofstede, R. (2011) 'Ecosystem-based marine spatial management: Review of concepts, policies, tools, and critical issues'. *Ocean and Coastal Management*, vol 54, no 11, pp. 807–820. doi:10.1016/j.ocecoaman.2011.09.002.

Kenworthy, W.J., Wyllie-Echeverria, S., Coles, R.G., Pergent, G. and Pergent-Martini, C. (2007) 'Seagrass conservation: An interdisciplinary science for protection of the seagrass biome', in A.W.D Larkum, R.J. Orth and C.M. Duarte (eds.), *Seagrass: Biology, Ecology and Conservation*, pp. 595–623. The Netherlands: Springer.

Kurihara, H., Kato, S. and Ishimatsu., A. (2007) 'Effects of increased seawater pCO_2 on early development of the oyster *Crassostrea gigas*'. *Aquatic Biology*, vol 1, no 1, pp. 91–98. doi:10.3354/ab00009.

Law, B.A., Hill, P.S., Maier, I., Milligan, T.G. and Page, F. (2014) 'Size, settling velocity and density of small suspended particles at an active salmon aquaculture site'. *Aquaculture Environment Interactions*, vol 6, no 1, pp. 29–42. doi:10.3354/aei00116.

Lin, D.T. and Bailey-Brock, J.H. (2008) 'Partial recovery of infaunal communities during a fallow period at an open-ocean aquaculture'. *Marine Ecology Progress Series*, vol 371, pp. 65–72. Available at: www.jstor.org/stable/24872686.

Lin, J., Li, C. and Zhang, S. (2016) 'Hydrodynamic effect of a large offshore mussel suspended aquaculture farm'. *Aquaculture*, vol 451, pp. 147–155. doi:10.1016/j.aquaculture.2015.08.039.

Lowe, R.J., Pivan, X., Falter, J., Symonds, G. and Gruber, R. (2016) 'Rising sea levels will reduce extreme temperature variations in tide-dominated reef habitats'. *Science Advances*, vol 2, no 8, p. e1600825. doi:10.1126/sciadv.1600825.

McLarney, W.O. (1984) *The Freshwater Aquaculture Book*. Point Roberts, WA: Hartley & Marks.

McNiven, I.J., Crouch, J., Richards, T., Dolby, N. and Jacobsen, G. (2012) 'Dating Aboriginal stone-walled fishtraps at Lake Condah, southeast Australia'. *Journal of Archaeological Science*, vol 39, no 2, pp. 268–286. doi:10.1016/j.jas.2011.09.007.

Meehl, G.A., Washington, W.M., Collins, W.D., Arblaster, J.M., Hu, A., Buja, L.E., Strand, W.G. and Teng, H. (2005) 'How much more global warming and sea-level rise?'. *Science*, vol 307, no 5716, pp. 1769–1772. doi:10.1126/science.1106663.

Morizane, T., Takimoto, S., Nishikawa, S., Matsuyama, N., Tyohno, K., Uemura, S., Fujita, Y., Yamashita, H., Kawakami, H., Koizumi, Y., Uchimura, Y. and Ichikawa, M. (2001) 'Mass mortalities of Japanese pearl oyster in Uwa Sea, Ehime in 1997–1999'. *Fish Pathology*, vol 36, no 4, pp. 207–216. doi:10.3147/jsfp. 36.207.

Najjar, R.G., Pyke, C.R., Adams, M.B., Breitburg, D., Hershner, C., Kemp, M., Howarth, R., Mulholland, M.R., Paolisso, M., Secor, D., Sellner, K., Denice, W. and Wood, R. (2010) 'Potential climate-change impacts on the Chesapeake Bay'. *Estuarine, Coastal and Shelf Science*, vol 86, no 1, pp. 1–20. doi:10.1016/j.ecss.2009.09.026.

Nash, C. (2010) *The History of Aquaculture*. Amsterdam: Wiley-Blackwell.

Nilsson, G.E., Dixson, D.L., Domenici, P., McCormick, M.I., Sørensen, C., Watson, S.A. and Munday, P.L. (2012) 'Near-future carbon dioxide levels alter fish behaviour by interfering with neurotransmitter function'. *Nature Climate Change*, vol 2, no 3, pp. 201–204. doi:10.1038/nclimate1352.

North, W.J. (1987) 'Biology of the Macrocystis resource in North America', in M.S. Doty, J.F. Caddy and B. Santelices (eds.), 'Case studies of seven commercial seaweed resources', pp. 265–311. FAO Fisheries technical paper 281.

O'Connor, W.A. and Lawler, N.F. (2004) 'Salinity and temperature tolerance of embryos and juveniles of the pearl oyster, *Pinctada imbricata* Röding'. *Aquaculture*, vol 229, no 1–4, pp. 493–506. doi:10.1016/S0044-8486(03)00400-9.

Páez-Osuna, F. (2001) 'The environmental impact of shrimp aquaculture: Causes, effects, and mitigating alternatives'. *Environmental Management*, vol 28, no 1, pp. 131–140. doi:10.1007/s002670010212.

Pahlow, M., van Oel, P.R., Mekonnen, M.M. and Hoekstra, A.Y. (2015) 'Increasing pressure on freshwater resources due to terrestrial feed ingredients for aquaculture production'. *Science of the Total Environment*, vol 536, pp. 847–857. doi:10.1016/j. scitotenv.2015.07.124.

Pereira, P.M., Black, K.D., McLusky, D.S. and Nickell, T.D. (2004) 'Recovery of sediments after cessation of marine fish farm production'. *Aquaculture*, vol 235, no 1–4, pp. 315–330. doi:10.1016/j.aquaculture.2003.12.023.

Pergent-Martini, C., Boudouresque, C.F., Pasqualini, V. and Pergent, G. (2006) 'Impact of fish farming facilities on *Posidonia oceanica* meadows: A review'. *Marine Ecology*, vol 27, no 4, pp. 310–319. doi:10.1111/j.1439-0485.2006.00122.x.

Pickering, T.D. (2006) 'Advances in seaweed aquaculture among Pacific Island countries'. *Journal of Applied Phycology*, vol 18, no 3–5, pp. 227–234. doi:10.1007/s10811-006-9022-1.

Poizot, E., Verjus, R., N'Guyen, H.Y., Angilella, J.R. and Méar, Y. (2016) 'Self-contamination of aquaculture cages in shallow water'. *Environmental Fluid Mechanics*, vol 16, no 4, pp. 793–805. doi:10.1007/s10652-016-9450-7.

Pouvreau, S., Bourles, K., Lefebvre, S., Gangnery, A. and Alunno-Brusci, M. (2006) 'Application of a dynamic energy budget model to the Pacific oyster, *Crassostrea gigas*, reared under various environmental conditions'. *Journal of Sea Research*, vol 56, no 2, pp. 156–167. doi:10.1016/j.seares.2006.03.007.

Purcell, S.W. and Simutoga, M. (2008) 'Spatio-temporal and size-dependent variation in the success of releasing cultured sea cucumbers in the wild'. *Reviews in Fisheries Science*, vol 16, no 1–3, pp. 204–214. doi:10.1080/10641260701686895.

Reid, G.K., Liutkus, M., Robinson, S.M.C., Chopin, T.R., Blair, T., Lander, T., Mullen, J., Page, F. and Moccia, R.D. (2009) 'A review of the biophysical properties of salmonid faeces: Implications for aquaculture waste dispersal models and integrated multi-trophic aquaculture'. *Aquaculture Research*, vol 40, no 3, pp. 257–273. doi:10.1111/j.1365-2109.2008.02065.x.

Ridler, N., Wowchuk, M., Robinson, B., Barrington, K., Chopin, T., Robinson, S., Page, F., Reid, G., Szemerda, M., Sewuster, J. and Boyne-Travis, S. (2007) 'Integrated Multi-trophic Aquaculture (IMTA): A potential strategic choice for farmers'. *Aquaculture Economics and Management*, vol 11, no 1, pp. 99–110. doi:10.1080/13657300701202767.

Salama, N.K. and Murray, A.G. (2011) 'Farm size as a factor in hydrodynamic transmission of pathogens in aquaculture fish production'. *Aquaculture Environment Interactions*, vol 2, no 1, pp. 61–74. doi:10.3354/aei00030.

Skladany, M., Clausen, R. and Belton, B. (2007) 'Offshore aquaculture: The frontier of redefining oceanic property'. *Society & Natural Resources*, vol 20, no 2, pp. 169–176. doi:10.1080/08941920601052453.

Snieszko, S.F. (1974) 'The effects of environmental stress on outbreaks of infectious diseases of fishes'. *Journal of Fish Biology*, vol 6, no 2, pp. 197–208. doi:10.1111/j.1095-8649.1974.tb04537.x.

Soma, K., Ramos, J., Bergh, Ø., Schulze, T., van Oostenbrugge, H., van Duijn, A.P., Buisman, E., Stenzenmüller, V., Grati, F., Mäkinen, T., Stenberg, C. and Kopke, K. (2013) 'The "mapping out" approach – Effectiveness of marine spatial management options in European coastal waters'. *ICES Journal of Marine Science*, vol 71, no 9, pp. 2630–2642. doi:10.1093/icesjms/fst193.

Soto, D., Aguilar-Manjarrez, J. and Hishamunda, N. (2008) 'Building an ecosystem approach to aquaculture'. FAO/Universitat de les Illes Balears Expert Workshop, 7–11 May 2007, Palma de Mallorca, Spain. FAO Fisheries and Aquaculture Proceedings 14, Food and Agriculture Organization of the United Nations, Rome.

Stentiford, G.D., Neil, D.M., Peeler, E.J., Shields, J.D., Small, H.J., Flegel, T.W., Vlak, J.M., Jones, B., Morado, F., Moss, S., Lotz, J., Bartholomay, L., Behringer, D.C., Hauton, C. and Lightner, D.V. (2012) 'Disease will limit future food supply from the global crustacean fishery and aquaculture sectors'. *Journal of Invertebrate Pathology*, vol 110, no 2, pp. 141–157. doi:10.1016/j.jip.2012.03.013.

Stevens, C., Plew, D., Hartstein, N. and Fredriksson, D. (2008) 'The physics of open-water shellfish aquaculture'. *Aquacultural Engineering*, vol 38, no 3, pp. 145–160. doi:10.1016/j.aquaeng.2008.01.006.

Tovar, A., Moreno, C., Mánuel-Vez, M.P. and García-Vargas, M. (2000) 'Environmental impacts of intensive aquaculture in marine waters'. *Water Research*, vol 34, no 1, pp. 334–342. doi:10.1016/S0043-1354(99)00102-5.

Viljugrein, H., Staalstrøm, A., Molvær, J., Urke, H.A. and Jansen, P.A. (2009) 'Integration of hydrodynamics into a statistical model on the spread of pancreas disease

(PD) in salmon farming'. *Diseases of Aquatic Organisms*, vol 88, no 1, pp. 35–44. doi:10.3354/dao02151.

Wainwright, T. (1982) 'Milkfish fry seasonality on Tarawa, Kiribati, its relationship to fry seasons elsewhere, and to sea surface temperatures (SST)'. *Aquaculture*, vol 26, no 3, pp. 265–271. doi:10.1016/0044-8486(82)90162-4.

Welladsen, H.M., Southgate, P.C. and Heimann, K. (2010) 'The effects of exposure to near-future levels of ocean acidification on shell characteristics of *Pinctada fucata* (Bivalvia: Pteriidae)'. *Molluscan Research*, vol 30, no 3, pp. 125–130. Available at: www.mapress.com/mr/content/v30/2010f/n3p130.pdf.

Young, C.N., Schopmeyer, S.A. and Lirman, D. (2012) 'A review of reef restoration and coral propagation using the threatened genus *Acropora* in the Caribbean and Western Atlantic'. *Bulletin of Marine Science*, vol 88, no 4, pp. 1075–1098. doi:10. 5343/bms.2011.1143.

11 Extreme human behaviours affecting marine resources and industries

Jade Lindley, Erika J. Techera and D. G. Webster

Introduction

Oceans have provided a number of resources and services of critical value to governments, communities and industries for centuries. The interest in extracting greater benefits from the oceans has led to the emergence of the blue economy agenda and goals. There is little doubt that commercial fishing, seabed mining, energy production, marine-based tourism and shipping all depend upon marine ecosystem services and resources, but if blue economy goals are to be based on or scaled up in these areas, good governance is essential to avoid negative impacts. Law can provide coherent and consistent frameworks within which legitimate activities can be efficiently and effectively undertaken. These frameworks depend, however, on adherence to the rules; damage can still be done if laws are poorly implemented and where enforcement is weak.

Illegal activities negatively impact upon the marine environment and, if unchecked, can also disincentivise legitimate users and operations. Illegal fishing, for example, is estimated to cost US$10–23.5 billion annually (Agnew *et al.*, 2009), the financial cost of forced labour is US$21 billion (ILO, 2009) and piracy between US$7 and US$12 billion per year (Bowden *et al.*, 2010). If blue economy goals are to be achieved then illegal behaviours, such as those discussed in this chapter, must be addressed. This necessarily involves regulating activities and prohibiting certain conduct, but also exploring and analysing motivators of crime and catalysts for compliance. This chapter examines crimes that occur in ocean areas: illegal fishing and human trafficking affecting the fisheries sector, and maritime piracy that has consequences for the shipping and transport industry. These crimes negatively impact upon legitimate marine-based activities that are the focus of blue economy goals, as well as the achievement of human rights.

As well as economic impacts, illegal, unreported and unregulated (IUU) fishing has been shown to have significant effects on marine living resources – depleting fish stocks and affecting ecosystems – the broader marine environment, the legitimate fishing industry and the livelihoods of fishers. If unchecked, IUU fishing will have negative environmental, economic and socio-cultural consequences affecting food and human security. Significant international, regional and national efforts have been made to sustainably manage fisheries and prohibit damaging behaviour; yet stock declines continue. The challenges of IUU fishing are not limited to preventing or addressing overfishing and overcapitalisation in fishing fleets. In its most extreme form, IUU fishing can extend beyond individual perpetrators to involve transnational organised crime on the high seas, with links to other maritime security issues including trafficking of people, weapons and drugs (Telesetsky, 2014). Therefore, identifying and addressing extreme behaviours is of critical importance.

In order to enhance fisheries governance, it is necessary to understand the circumstances that facilitate ongoing regulatory breaches and illicit activities. This chapter explores the issue of human trafficking in this context. Use of forced labour is increasing in fisheries around the world, and while this is often seen as primarily a law enforcement problem, it is also a fisheries governance issue. Human trafficking lowers operating costs, which in turn increases incentives to expand fishing effort. Trafficking enables connections between fishers and organised crime, further facilitating illegal fishing through the availability of smuggling networks and access to syndicate resources. This chapter reviews what is known about forced labour in the fishing industry and shows how the same factors that contribute to the persistence of panaceas also enable the trafficking to continue.

Extreme human behaviour can also be seen in terms of violence and robbery at sea, most commonly in the context of piracy. Piracy has a significant global impact on the shipping industry: there are financial costs associated with ransoms and enhanced security overheads, and human costs through death and injury to merchant crews. While incidents of piracy (predominantly Somali piracy in the Indian Ocean) reduced following a range of international efforts, there has been a recent resurgence. This extreme and illegal behaviour has appeared to defy interventions and international legal responses. This chapter explores these issues with a focus on the reasons people turn to piracy; in many cases these are linked to loss of livelihoods and environmental degradation.

The chapter explores three areas – illegal fishing, human trafficking and maritime piracy – which each confound legal frameworks. The underlying causes and facilitators are examined as well as responses to the challenges. Importantly, the interlinkages and implications for blue economy goals are analysed. The chapter concludes by examining ways forward to ensure extreme human behaviours are minimised.

Illegal fishing

Fishing is one of the oldest human activities associated with our oceans. Coastal communities have caught fish for food for millennia, and even trade in fish has a long-standing history. While populations were low impacts were minimal, but over time rules and regulations were developed to manage fishers and fishing. Although the specific detail changes across the world, these rules have been fairly uniform in approach: controlling the number of fishers and the amount of fish caught, restricting which species may be taken, limiting where and when fish may be caught, as well as managing the gear and equipment used (Webster, 2015). These rules have been utilised in international, regional and national legal frameworks for sustainable fisheries regulation and management. Legitimate fishers operate within these established frameworks of rules; illegal fishers do not.

Although overfishing was first identified as a problem in the 1800s, it was in the latter half of the twentieth century that growing concerns were raised about the impacts of fishing. Towards the end of the century, the first global efforts were made to address illegal fishing alongside unregulated and unreported activities. The *Convention on the Conservation of Antarctic Marine Living Resources*, for example, made reference to illegal fishing in the 1990s, and the Food and Agriculture Organization (FAO) subsequently acknowledged IUU fishing as a global problem in 2001 (Baird, 2004). IUU fishing is a problem that plagues most oceans and seas including various sectors of the Pacific, Indian and Atlantic Oceans as well as the Arufura and Bering Seas. Illegal fishing focuses upon the offender, as does unreported fishing to a great extent. Unregulated fishing places emphasis on the government or entity with the obligation or authority to manage a fishery or fishers. In combination, illegal, unregulated and unreported fishing has been recognised as a significant threat to fish stocks and sustainable ocean management (Telesetsky, 2014; Lindley and Techera, 2017). Efforts have been made to address regulatory gaps and, although some concerns remain, the major challenge is implementation at the national level as well as monitoring and enforcement of rules.

Implementation of fisheries regulations relies upon government will and national resources at the domestic level to operationalise and enforce fisheries management laws. Monitoring legitimate fishers involves a combination of self-reporting, on-board fisheries observers, port-based measures and vessel tracking using vessel monitoring systems and automatic identification systems (VMS and AIS) as well as satellites. Surveillance for illegal activities relies upon police, coastguard or naval vessels utilised for constabulary services. Ultimately, enforcement involves fines for infringements or prosecutions for fishery crimes; again, these are dependent upon political will, strong governance and domestic resources to achieve outcomes. Where governance is weak, or corruption is widespread, outcomes are impeded. Similarly, government regulation of fisheries requires legitimacy if it is to be successful. At its simplest, illegal fishing may involve small scale fishers taking a few more fish than their quota or bag limit allows. Alternatively, it may include using prohibited fishing equipment (dynamite or banned net types) or methods (such as live shark finning or failure to use bycatch reduction gear), or straying into a fishing-restricted area. While individual infringements may not have a significant impact, where they become common and widespread their cumulative impacts can be great.

In areas close to shore, including the territorial sea (0–12 nautical miles), national government agencies have strong powers to enforce fisheries regulations against both domestic and foreign fishers. Although such enforcement effort involves the use of State resources, fishing is more visible in these areas, often dominated by domestic fleets, and the majority of catches are landed at ports and sold in local markets. Thus, fishing in areas close to shore is theoretically simpler to manage and illegal fishing is easier to identify and address. Where there are large numbers of small vessels, however, this can be a monitoring challenge, and cumulative impacts on fish stocks can be significant. With very large commercial operations, particularly those involving activities further from the shore, the challenge becomes harder. Large-scale operations have the potential to make a more significant impact and, where entire illicit fishing fleets or transnational criminal operations are involved, extreme effects on the marine environment can include decimation of stocks, ecosystems and habitats. The further offshore, the more widespread the activities and the more sophisticated the operations, the harder the issue is to address as vessels may be flagged in different countries, fishers may also be drawn from various countries and their activities may cross different national borders. When fish are transferred between vessels at sea (transshipped), rather than brought back to a port for offloading, this further complicates reporting on catches. Where fishing occurs on the high seas, a great distance from land and in areas beyond national jurisdictions, surveillance is particularly difficult. The international community has sought to address the myriad challenges through various instruments and organisations, as set out below, and in particular to assist developing countries which have limited resources to combat illegal activities.

The *UN Convention on the Law of the Sea* was adopted in 1982. Although it includes provisions relating to fisheries and marine environmental protection, it does not mention illegal fishing or fisheries crime. The 1995 *UN Fish Stocks Agreement* creates flag State obligations to ensure compliance with regional conservation and management measures, and to take enforcement measures including the imposition of sanctions. At the international level the FAO has taken the lead, setting standards for sustainable fishing and addressing IUU fishing within its mandate to ensure food security. In 1995, the FAO developed the *Code of Conduct for Responsible Fisheries*, which sets out general measures for sustainable fisheries development. The FAO has specifically sought to address illegal fishing through its 2001 International Plan of Action on IUU Fishing (IPOA-IUU). This is a non-binding instrument, but influential because it provides a standardised framework for domestic measures and, for example, requires flag states to ensure vessels are authorised to fish in waters beyond their jurisdiction. The 2009 *FAO Agreement on Port State Measures to Prevent, Deter and Eliminate Illegal, Unreported and Unregulated Fishing* focuses on the point at which fish are landed and compliance with conservation and fisheries regulations. It permits authorities to deny landings and refuse permission to dock if they suspect IUU fishing, even by foreign fishing vessels, encourages harmonisation of port State measures, enhances regional and international cooperation and seeks to block the flow of IUU-caught fish into markets.

At the regional level efforts have been made through inter-governmental organisations to address monitoring and enforcement issues. In the Indian Ocean, for example, the Indian Ocean Commission (IOC) has facilitated cooperative measures among member States to share data on illegal vessels, activities and corporations through the Regional Programme for Fisheries Surveillance and the IOC SmartFish program (IOC, 2018). In the Pacific the States have gone further, adopting the *Niue Treaty on Cooperation in Fisheries Surveillance and Law Enforcement in the South Pacific Region* (Niue Treaty, 2013). Regional Fishery Management Organisations (RFMOs) have also sought to address the problem. For example, the Indian Ocean Tuna Commission keeps a log of illegal fishing vessels known to operate in the region (IOTC, 2018a) and has prepared reports analysing illegal fishing (IOTC, 2018b). In the Atlantic the North East Atlantic Fisheries Commission requires all vessels fishing in the region to comply with current management measures and its Scheme of Control and Enforcement, otherwise they are treated as engaging in IUU fishing (NEAFC, 2018). The US has also assisted many Pacific countries through its ship-rider agreements whereby foreign enforcement officers can travel on board US Coast Guard vessels, thereby enhancing monitoring efforts (CSIS, 2014).

The above measures have been taken to address IUU fishing through fisheries regulations and institutions. Yet, it has become clear that in many cases other factors facilitate the persistence of fisheries crime, and are indeed crimes themselves, including the use of forced labour discussed below.

Human trafficking and the fishing industry

Although human trafficking and forced labour are not new, the extent of these practices in modern fisheries only came to light in the early to mid-2000s. Human trafficking involves moving people from one location to another, whether by coercion or fraud, in order to exploit them in some way, usually for prostitution or forced labour. Means of forcing people to work include debt bondage, holding immigration papers or threatening to report illegal migrants to authorities, verbal and physical abuse and physical barriers to prevent people from leaving the workplace. While illegal migrants are particularly vulnerable to both debt bondage and threats that they will be reported to the authorities, legal migrants and minorities from within a country can also be exploited in similar ways.

There are relatively few global studies of human trafficking and forced labour in the fishing industry, but research has found exploitation of both migrant and non-migrant populations on board fishing vessels, in aquaculture ponds and in processing facilities worldwide. Human rights violations have been noted in the fisheries sectors of a number of developing countries, including India, China, the Philippines, Cambodia, South Africa and Thailand, where estimates suggest that tens of thousands of migrants and minorities are forced to work in all sectors of the fishing industry (Ratner *et al.*, 2014). Evidence suggests that in Bangladesh, children are forced to work in the fish processing industry on the islands of Dublar Char (Jensen, 2013). However, this is not just a 'developing country' problem; there are a number of reports of trafficking to provide fleets in the US, the UK and other developed countries with cheap labour (Ratner *et al.*, 2014). Exploitation of migrant labour has also been documented on board fishing vessels in New Zealand. Initially identified in the 1990s, this issue again rose to public attention in 2011 when the Indonesian crew of a South Korean vessel, the *Shin Ji*, fled the vessel in a New Zealand port, claiming long-standing abuse and exploitation. Since that time, more crews have come forward and media coverage of the issue continues (Simmons and Stringer, 2014; Stringer and Simmons, 2015; Stringer *et al.*, 2016). Recent news reports point out similar practices within the UK fleet, particularly the manipulation of migrant workers from countries such as the Philippines, who are allegedly brought into the UK illegally and forced to work very long hours for much lower wages than their European counterparts (Lawrence *et al.*, 2015). The same types of abuses of migrant labour have been found in the US as well (Mendoza and Mason, 2016). Indeed, human trafficking has been found in all sectors of the fishing industry, all around the world (ILO, 2013; Lewis *et al.*, 2017; EJF, 2018).

There are many legal, economic and political factors that contribute to the prevalence of human trafficking and forced labour in fisheries. Economically, as fish stocks decline due to overfishing, profit margins narrow in all segments of the industry. This may be most obvious on fishing boats, which have to increase effort to catch the same or fewer fish. Some of these higher costs will be passed on to processors who buy their raw materials from fishers. In aquaculture, the problem is somewhat different. Rather than overexploiting fish, culturalists tend to overexploit areas available for fish production (especially for high quantity, low cost products like shrimp) forcing up operating costs even as prices are driven down by increasing overall supply of cultured seafood (Biao and Kaijin, 2007; Lebel *et al.*, 2010; Verité, 2015). Since labour is usually a large component of operating costs for all of these sectors, there are strong incentives for producers to find low cost sources of labour. Furthermore, by reducing operating costs, forced labour actually increases the potential for overexploitation of natural resources by allowing fishing vessels, processors or aquaculture organisations to stay in business longer. This creates a positive feedback between human trafficking and overexploitation (of either fish stocks or aquaculture resources) that makes each problem worse than it would be in isolation.

There are similar feedbacks in the economics of organised crime, human trafficking and IUU fishing. In the last few decades, high levels of connectivity via the Dark Net have made it much easier and cheaper for criminal gangs to organise trafficking activities of all kinds. It is not clear how these networks extended into the fishing industry, but once a few connections were made others followed, ultimately invading many sectors. In other words, as producers entered the underground economy, whether by buying IUU fish or by employing trafficked workers, it became easier to move into related criminal activities. A processor might introduce a fishing boat captain to a trafficker of labour or vice versa. Furthermore, for fishers, processors or culturalists working in a particular region, increasing competition from other producers using forced labour increases the incentives to use trafficking victims, even while making it easier to forge the connections needed to secure that labour. This process can generate a race to the bottom that rapidly entangles the fishing industry and organised crime in a given area (Albanese, 2011; Bondaroff *et al.*, 2015, p. 84).

Politically, human trafficking and forced labour are facilitated by: (1) corruption and lack of enforcement capacity; (2) prejudices against migrants and other marginalised peoples; and (3) the complexity of seafood supply chains, which makes it difficult to trace connections between products and suppliers. First, corruption and lack of capacity are most important in developing countries, which are currently the primary locations for processing and aquaculture facilities; but they are also important in developed countries, where operating costs are higher and so pressures to keep labour costs low are greater. Distant water fishing fleets pose a particular problem in that they are highly mobile and may not return to land for months or even years at a time. Many argue that banning transshipment at sea and increasing the ability of port States to search and seize vessels from other countries could help to reduce this problem by increasing the chance that trafficking will be detected. This has led to international legal developments such as the *FAO Agreement on Port State Measures to Prevent, Deter and Eliminate Illegal, Unreported and Unregulated Fishing*, discussed above, as well as the International Labour Organization (ILO) *Maritime Labour Convention* and *Convention Concerning Work in the Fishing Sector* (Lindley and Techera, 2017). Such laws alone are unlikely to eradicate the problem as evidence indicates some captains will strand their crew on rocks or deserted islands before the vessel lands in port, then pick them up again on their way back out to sea (Crane, 2013; Surtees, 2013). Therefore, additional measures are needed to curb these practices, along with improved domestic-level monitoring and enforcement for inshore and coastal segments of the industry.

Second, a lack of political will, which may be tied to prejudices against migrants or domestic minorities, makes it difficult to implement measures that target human trafficking. Many authors have shown that the majority of trafficked people in the fishing industry were not citizens of the country in which they were enslaved or, in some cases, they were ethnic minorities that were forced to migrate from rural to urban areas in search of work (Bales, 2012; ILO, 2016, 2017; Lewis *et al.*, 2017). Both internal and external migrants were politically marginalised by their lack of citizenship or equivalent political standing, and socially marginalised by prejudice against outsiders or minority populations. This occurred in aquaculture operations, terrestrial processing facilities and on fishing vessels. The only observed counter-trends occurred when public attention was focussed on the issue, threatening the brand identities of large corporations with considerable economic clout. In some cases, the trafficked workers escaped and were able to bring attention to their cause. In other cases, consumers, scholars and reporters brought greater attention to the issue, often with the help of non-governmental organisations (NGOs). However, this scrutiny tends to be sporadic and has only been sustained in the case of Thailand, which has faced considerable international pressure to curb human trafficking in its fishing industry in recent years (Derks, 2010; Bales, 2012; Legacy Phuket Gazette, 2014; Chantavanich *et al.*, 2016; Kelly, 2016; Lawrence and Hodal, 2017).

Third, the complexity of global supply chains makes the international interventions described above much more difficult. As the industry has globalised over the last 70 years, supply chains became long and convoluted, making it much more difficult to trace products and much easier to hide human trafficking, forced labour and other transgressions. Consumers may not even know where their fisheries products come from, let alone whether or not they are harvested and processed in a sustainable and humane manner. For example, in 2015, three law firms sued Costco and their Thai seafood supplier, Charoen Pokphand (CP) Foods, for knowingly selling shrimp associated with human rights abuse and violating the California *Transparency in Supply Chains Act 2010* (Larson, 2015). A year later a lawsuit was filed against Nestle on similar grounds. Though it won the lawsuit,[1] Nestle admitted that it could not in fact trace its shrimp supply chain and therefore could not verify that no slave labour was used in the process. They therefore commissioned an NGO called Verite to study the issue. The research showed that violations for land-based fisheries production in Thailand were similar to sea-based production, and that the same types of coercion were used to control workers on shrimp farms and in processing facilities (Verité, 2015, p. 26). Labelling programmes such as the Marine Stewardship Council and FishWise, which were developed to verify the sustainability of fisheries, are also considering expansion to include human trafficking, but this has proved to be quite difficult.

The combined forces of large corporations and NGOs may be a powerful tool in the fight against forced labour in fisheries, but international agencies are also stepping up. In addition to the Port State Measures Agreement developed by the FAO, INTERPOL is coordinating efforts to combat both IUU and human trafficking in fisheries through its Project Scale initiative. Although not designed to tackle forced labour specifically, the EU's IUU Card programme, which allows the EU to sanction countries found to have high levels of IUU activity, is frequently cited as one of the reasons for recent improvements in Thailand's regulation of human trafficking in its fishing industry. The US has also used its *Trafficking Victims Protection Act* to both document and threaten Thailand with sanctions if it does not curb human trafficking. Though results are slower than NGOs would like, it is this type of combined pressure that can really help to reduce the problem of forced labour in fisheries.

Unfortunately, international politics also plays a role here, as the US and EU are unlikely to pressurise other powerful countries with documented human trafficking problems in their fisheries, such as China and South Korea, or within their own nations. This further highlights the importance of increasing supply chain transparency; educating people about the supply chain issue is important in developing international political will to act. If standards are not improved at the global scale traffickers and their industry partners may simply move to places where governance and transparency are weaker. Much like the IUU fishing issue itself, fisheries related forced labour and human trafficking are moving away from the spotlight in Thailand to other countries such as India and Taiwan (Sutton and Siciliano, 2016, p. 32; Lewis *et al.*, 2017; EJF, 2018). Nevertheless, increasing news coverage and NGO and business concern, as well as international programmes such as Project Scale, are heartening developments.

The illegal fishing and forced labour issues discussed above do not exist in a vacuum. There are other maritime crimes that in some cases have links to fishing and in others operate independently. Regardless of the connections between them, these extreme human behaviours together place pressure on the ocean environment and the governments tasked with enforcing the law, as well as the achievement of blue economy goals.

Maritime piracy

Maritime piracy is regarded as the oldest international crime and sparked the first internationally harmonised response, yet global piracy control remains an elusive goal (United Nations Secretary-General, 2010). The earliest documented incidents of piracy date back to the 1300s BC, although the better-known golden age of piracy occurred in the sixteenth and seventeenth century Caribbean. Piracy was outlawed early on by most seafaring nations and largely eradicated in the nineteenth century; nevertheless, it was considered so serious that the *United Nations Convention on the Law of the Sea* (UNCLOS) gives universal jurisdiction to all States to address it where it occurs on the high seas (UNCLOS, 1982). Despite these interventions a new age of maritime piracy has developed and persisted.

In order to understand piracy's maritime extreme, it is important to reflect on its contemporary nature and extent and the instruments in place to respond to it. Maritime piracy has manifested in distinct ways in various parts of the world and has been regarded differently; as such the control of piracy must also be approached with a region specific view. By the end of 2017, globally, pirates had attacked 180 commercial and private vessels

(International Maritime Bureau, 2018). This compares to 2012 when Somali piracy was at its peak of attacks and 445 attacks were recorded globally (International Maritime Bureau, 2013). While in 2000, the year of the highest number of reported attacks in the past three decades, 469 attacks occurred, predominantly due to Indonesian pirates (International Maritime Bureau, 2003).

In addition to the number of attacks, piracy trends have changed over the past 30 years, likely a reflection of changing geo-political relationships as well as changing internal governance capabilities and defence strategies in known piracy hotspots. During this period, piracy occurred around four main centres: South-East Asia, the Indian Sub-Continent, East Africa and West Africa (International Maritime Bureau, 2018). Unsurprisingly, the hotspots are located around weak/failing States with high levels of corruption or limited law enforcement on land, or with limited naval forces or coastguards in port or at sea to protect passer-by vessels. While the South-East Asia and Indian Sub-Continent hotspots have historical connections to piracy, pirates operating in the two African hotspots are opportunists. Due to the existence of both historical 'culturally-normalised' robbery at sea and the contemporary (re)emergence of maritime piracy, it is foreseeable that piracy will continue worldwide to some extent. However, as the threat of piracy in various high risk zones differs, a one-size-fits-all approach is unlikely to be successful; therefore successful responses are dependent upon actions by both regional neighbours and global stakeholders.

Low levels of piracy exist around the world, though sustainable regional actions reduce the need for expensive global interventions. In South-East Asia, piracy has historical origins and is unlikely to disappear completely, despite strong commitment to control it within that region. Piracy around Africa would have likely existed before, to a much smaller extent than seen in recent years, due to the existence of commercial vessels transiting near land that utilises long-standing sea routes. In recent years, the permeation of piracy around Africa's west and east coasts has been comparatively significant and it is likely to continue, albeit at a relatively minor rate. Therefore, in devising suitable and long-term sustainable responses, it is likely that the international community must accept a modest level of piracy.

Motivations likely differ between traditional and non-traditional pirates. Traditional pirates use stealth and minimal force to take advantage of vulnerable vessels near shore. For instance, the Malacca Strait, a known chokepoint, slows vessels entering the Strait and the ports of Singapore. This makes it easy for pirates to board vessels and steal supplies, often without crews' knowledge (International Maritime Bureau, 2013). Similarly, the chokepoint into Chittagong port allows Bangladeshi pirates to offend on vessels nearing to anchor, most commonly armed with knives (International Maritime Bureau, 2018). This contrasts to non-traditional pirates of the west and east coasts of Africa who conduct much riskier attacks, armed with semi-automatic weapons and targeting steaming vessels for the purpose of stealing cargo and taking crew and vessels hostage (International Maritime Bureau, 2018). Continuous efforts to reduce pirates' motivation, decrease opportunity through awareness and prevention and increase reliable and capable guardianship could succeed in reducing the problem.

The responses to piracy have been many and varied, according to the regional need. Somali piracy yielded the greatest global buy-in due to the number of commercial vessels transiting the high seas in the region, meaning that a large majority of States were affected, whether littoral or landlocked. Due to the work of the United Nations in prompting States to codify modern counter-piracy legislation there is now, promisingly, improved legal homogeneity across borders, though penalties may differ (Lindley, 2016, p. 80).

In 2016, in response to Somali and Nigerian piracy and related issues, the African Union (AU) adopted a regional agreement, the African Charter on Maritime Security, Safety and Development in Africa (the Lomé Charter). Well received by AU Member States, 35 of 55 States signed the binding agreement during the assembly session (Organization of the African Unity, 2018). The purpose of the agreement is to increase maritime security and act on maritime crimes to protect and secure the AU seas and oceans, including from piracy (Organization of the African Unity, 1994). It emphasises the need to strengthen training and professionalisation of navies, coastguards and other agencies responsible for border and maritime security (Organization of the African Unity, 1994). The agreement acknowledges the limitations of some of its Members, requesting States to endeavour to develop systems of information sharing (Organization of the African Unity, 1994). Timely and transparent information and data transfer between stakeholders is believed to be one of the greatest weapons against piracy (International Chamber of Commerce, 2017). The lack of urgency afforded to information exchange may inhibit the response to piracy. Further, the success of the *Lomé Charter* hinges on the adoption of the 1994 *African Maritime Charter* and its 2010 revised treaty; neither have entered into force (Organization of the African Unity, 1994). Given the ongoing nature and extent of piracy around Africa, the urgency of adopting and implementing these instruments cannot be overstated.

In addition to responses to piracy around the Horn of Africa, piracy responses exist elsewhere such as South-East Asia, led by the Association of Southeast Asian Nations (ASEAN). ASEAN's regional response to piracy builds on the international response, encouraging Member States to adopt relevant international binding instruments and codify counter-piracy laws domestically. Once they have the necessary legal frameworks in place to enable prosecution upon interception, the response calls on Member States to establish greater protection measures in readiness of piracy and other threats to security (ASEAN, 2003). These protection measures include identifying and patrolling vessel traffic lanes; information sharing among Member States through bi- and multi-lateral arrangements; encouraging reporting of attacks and threats to regional reporting centres; and encouraging training and on-board vessel protection to minimise the likelihood of attacks (ASEAN, 2003).

ASEAN Member States cover the vast Asia Pacific region east and west between the US and Bangladesh, and north to south from Japan to Australia. The coverage of ASEAN is extensive and, therefore, so is the response to piracy and other security threats.

Other regions at risk of future or emerging piracy outside ASEAN require localised responses. While over-romanticised (thanks to Hollywood), the Caribbean region faces ongoing, albeit low threats of piracy. The Association of the Caribbean States has not specifically established a regional response to piracy. Compared to other regions, the threat is presently low with the few attacks reported rarely resulting in violence towards or harm of crew; incidents tend to occur at anchorage and involve minimal payouts where vessels are accosted (Lusk, 2016; International Maritime Bureau, 2018). The minor nature and extent of piracy in the Caribbean may also affect reporting. The involvement of law enforcement and insurance is costly and time-consuming compared to the harm caused, deterring some owners, especially those of private pleasure craft, from reporting incidents. Nevertheless, as piracy is a persistent challenge in other regions, the potential for increased frequency and seriousness of attacks in the Caribbean is not unforeseeable and a focussed regional response should be developed. This is particularly important given the existence of other transnational marine-based crimes operating from and transiting within the Caribbean region, such as the drugs trade. Furthermore, increasingly tighter controls of piracy in other regions may expose the Caribbean region. This was the case, for example, when naval vessels relocated to patrol the Indian Ocean, resulting in reduced surveillance around Indonesian waters and an unexpected piracy resurgence (International Maritime Bureau, 2018). Proactive planning may, therefore, be appropriate.

Piracy is an extreme and violent form of illegal maritime behaviour. Concerted international efforts yielded positive results off the east coast of Africa, but as security efforts are wound down, the threat of piracy is re-emerging. Economic, political, geographical as well as cultural issues are at play, but it also prompts an enquiry into underlying motivators and facilitators and the interlinkages of piracy with other crimes that occur on our oceans.

Interlinkages

The three areas of extreme human behaviours in the maritime realm discussed above involve illegal activities for economic gain. Each of these extremes impacts on the achievement of blue economy goals: illegal fishing and forced labour on board vessels undermine sustainable fisheries management and human rights, and may also disincentivise legitimate fishers from complying with the law. Piracy affects the shipping and transport sector and, indirectly, international trade itself. The three areas invoke complex issues, and solutions require an understanding of the underlying drivers, as well as reducing facilitating factors such as weak governance, corruption and poor enforcement.

These three areas of criminal activity are not independent of each other and research has demonstrated the interlinkages between organisations and individuals involved in various forms of extreme human behaviours. Significant global legal efforts have been made to address IUU fishing through standard setting for sustainable fishing, data sharing and improved monitoring and enforcement, as well as measures such as awareness raising, education and training. As noted above, the issue of forced labour in the fishing sector has also received global legal attention, but further international coordination, national endorsement and domestic legal provisions are needed as well as education across multiple sectors and communities. The issue of piracy was brought to global attention in the last decade. International legal efforts, beyond criminal prohibitions, saw the utilisation of international collaborations to address the Somali piracy issue. Yet these efforts will not be successful if those involved in illicit activities have no other livelihood options or where IUU fishing is facilitated by low cost inputs.

It has been shown, for example, that former fishers turned to piracy when their traditional livelihoods were lost. This was certainly true for Somalia after the demise of the previously thriving fishing industry in the 1990s, though the success of the Somali piracy business model inspired pirates from other regions to adopt a similar hijack-for-ransom model (International Maritime Bureau, 2018, p. 12). Both maritime piracy and human trafficking pose significant security and human threats and while offenders may not engage in both illegal activities, organised criminal groups that finance their activities may have implicit links to both. However, while the end goal of pirates and human traffickers align, in that both seek profit, their modus operandi differs significantly.

Evidence does not suggest that pirates ordinarily engage in human trafficking. However, probable links between piracy and human smuggling exist. Indeed, smuggling people the short distance to escape from atrocities in Ethiopia and Somalia to refugee camps in Yemen and Saudi Arabia is well known (United Nations Office on Drugs and Crime, 2013, p. 11). In South-East Asia, hijacked and rebirthed vessels, which provide some level of anonymity, may be used to traffic people, drugs and weapons, more closely linking pirates and traffickers (Emmers, 2003, p. 8). What is more likely, however, is the potential for people being smuggled to become 'extremely vulnerable to human trafficking, abuse, and other crimes', given that they are 'illegally present in the country of destination and often owe large debts to their smugglers' (United States Department of State, 2017). Once trafficked, those victims may be forced to work on IUU fishing vessels, for example.

In order to address the interlinked problems of IUU fishing and forced labour, the legal frameworks and institutions must work more closely together with coordinated efforts and responses. Although the *ILO Convention Concerning Work in the Fishing Sector* has been adopted it has not yet come into force; further signatories are needed and national legal protections must be put in place. It is in this area that lessons may be learned from global efforts to address Somali piracy. The international taskforce approach plus the efforts made to encourage national adoption of legislation to facilitate prosecutions could be followed to address IUU fishing and force labour issues.

The supply chain issue is one that has yet to be successfully tackled at the international level. The issue is not limited to fish as attention is periodically brought to other food, clothing and products that have suspect supply chains involving forced and 'slave' labour in their production (Walk Free, 2017). This lends weight to the need for global efforts to homogenise labelling systems and other mechanisms to raise consumer and supplier awareness. The globally recognised Marine Stewardship Council (MSC) recognises sustainable fisheries and could incorporate labour practices in the fishing industry as criteria for certification. International law could play a standard-setting role, incorporate non-state actors such as the MSC and provide a mechanism for the sharing of innovative domestic measures, such as the *Transparency in Supply Chains Act* noted above.

Conclusion

This chapter has highlighted three examples of illegality occurring on our oceans. While each is significant in itself, causing environmental, economic and social damage, where they overlap or the effects are cumulative the extreme conduct can have devastating consequences, particularly in circumstances where there is a widening focus on the blue economy. Good governance is key to addressing these extreme human behaviours, but so too is awareness raising across the broader community as well as specific industries and actors in marine resource and maritime transport supply chains.

Given the freedom that exists on the high seas, this chapter reconfirms the need for collective regional and international responses to minimise fragmentation and ensure effective harmonisation in responses, where possible. While there exists a plethora of suitable domestic and international legal frameworks and bi- and multi-lateral arrangements, pooling resources to respond more effectively is needed to maximise effort and minimise cost.

Furthermore, this chapter brings sharply into focus the need to better understand and address the drivers motivating these extreme human behaviours, such as limited legitimate employment opportunities, weak law enforcement and corruption at all levels of government. Addressing these peripheral factors may indirectly have a significant impact on the viability of these crimes.

Note

1 Both cases were ultimately dismissed as California law only requires companies to provide accurate information about the steps they are taking to monitor their supply chains and does not require firms to certify that the product was produced ethically (Rosenblatt, 2017).

References

Agnew, D. J., Pearce, J., Pramod, G., Peatman, T., Watson, R., Beddington, J. R. and Pitcher, T. J. (2009) 'Estimating the worldwide extent of illegal Fishing'. *PLoS ONE*, vol 4, no 2, p. e4570. doi:10.1371/journal.pone.0004570.

Albanese, J. S. (2011) *Transnational Crime and the 21st Century*. New York: Oxford University Press.

ASEAN (2003) 'ASEAN Regional Forum (ARF) statement on cooperation against piracy and other threats to security'. Association of Southeast Asian Nations (ASEAN), Singapore. Available at: http://asean.org/arf-statement-on-cooperation-against-piracy-and-other-threats-to-security/ (accessed 5 April 2018).

Baird, R. (2004) 'Illegal, unreported and unregulated fishing: An analysis of the legal, economic and historical factors relevant to its development and persistence'. *Melbourne Journal of International Law*, vol 5, no 2, pp. 299–334.

Bales, K. (2012) *Disposable People; New Slavery in the Global Economy*. Berkeley, CA: University of California Press.

Biao, X. and Kaijin, Y. (2007) 'Shrimp farming in China: Operating characteristics, environmental impact and perspectives'. *Ocean and Coastal Management*, vol 50, no 7, pp. 538–550. doi:10.1016/j.ocecoaman.2007.02.006.

Bondaroff, T. N. P., Van Der Werf, W. and Reitano, T. (2015) 'The illegal fishing and organized crime nexus: Illegal fishing as transnational organized crime'. The Global Initiative, Geneva. Available at: http://globalinitiative.net/wp-content/uploads/2015/04/the-illegal-fishing-and-organised-crime-nexus-1.pdf (accessed 13 November 2018).

Bowden, A., Hulburt, K., Aloyo, E., Marts, C. and Lee, A. (2010) 'The economic cost of maritime piracy'. One Earth Future Working Paper, December 2010. Available at: http://oceansbeyondpiracy.org/sites/default/files/attachments/The%20Economic%20Cost%20of%20Piracy%20Full%20Report.pdf (accessed 2 November 2018).

Chantavanich, S., Laodumrongchai, S. and Stringer, C. (2016) 'Under the shadow: Forced labour among sea fishers in Thailand'. *Marine Policy*, vol 68, pp. 1–7. doi:10.1016/j.marpol.2015.12.015.

Crane, A. (2013) 'Modern slavery as a management practice: Exploring the conditions and capabilities for human exploitation'. *Academy of Management Review*, vol 38, no 1, pp. 49–69. doi:10.5465/amr.2011.0145.

CSIS (2014) 'The value of ship rider agreements in the pacific. Center for Strategic & International Studies. Available at: www.cogitasia.com/the-value-of-ship-rider-agreements-in-the-pacific/ (accessed 20 April 2018).

Derks, A. (2010) 'Migrant labour and the politics of immobilisation: Cambodian fishermen in Thailand'. *Asian Journal of Social Science*, vol 38, pp. 915–932. doi:10.1163/156853110X530804.

EJF (2018) 'Human trafficking in Taiwan's fisheries sector'. Environmental Justice Foundation. Available at: https://ejfoundation.org/resources/downloads/Human-trafficking-in-Taiwan%E2%80%99s-fisheries-sector.pdf (accessed 8 May 2018).

Emmers, R. (2003) 'The threat of transnational crime in Southeast Asia: Drug trafficking, human smuggling and trafficking, and sea piracy'. Institute of Defence and Strategic Studies, Singapore. Available at: www.redalyc.org/html/767/76711296006/ (accessed 5 April 2018).

ILO (2009) 'Cost of coercion'. International Labour Organization. Available at: www.ilo.org/wcmsp5/groups/public/@dgreports/@dcomm/documents/generic-document/wcms_106200.pdf (accessed 11 June 2018).

ILO (2013) 'Caught at sea: Forced labour and trafficking'. International Labour Organization. Available at: www.ilo.org/wcmsp5/groups/public/-ed_norm/-declaration/documents/publication/wcms_214472.pdf (accessed 16 February 2017).

ILO (2016) 'Forced labour, modern slavery and human trafficking'. International Labour Organization. Available at: www.ilo.org/global/topics/forced-labour/lang-en/index.htm (accessed 16 February 2017).

ILO (2017) 'Statistics on forced labour, modern slavery and human trafficking'. International Labour Organization. Available at: www.ilo.org/global/topics/forced-labour/statistics/lang--en/index.htm (accessed 8 January 2017).

International Chamber of Commerce (2017) '4 takeaways from the IMB's latest global piracy report'. ICC news report, 17 October 2017. Available at: https://iccwbo.org/media-wall/news-speeches/4-takeaways-imbs-latest-global-piracy-report/ (accessed 5 April 2018).

International Maritime Bureau (2003) 'Piracy and armed robbery against ships: Annual report 1 January–31 December 2002'. International Chamber of Commerce, Essex.

International Maritime Bureau (2013) 'Piracy and armed robbery against ships: Annual report 1 January–31 December 2012'. International Chamber of Commerce, London.

International Maritime Bureau (2018) 'Piracy and armed robbery against ships: Annual report 1 January–31 December 2017'. International Chamber of Commerce, London.

IOC (2018) 'The Fisheries Ministerial Conference South-West Indian Ocean: United for a shared prosperity through the Blue Economy PRSP: Championing the battle against IUU Fishing', 18–21 July 2017, International Conference Centre, Antananarivo, Madagascar.

IOTC (2018a) 'Vessels'. Available at: www.iotc.org/vessels (accessed 20 January 2018).

IOTC (2018b) 'Report on presumed IUU fishing activities in the EEZ of Somalia'. Indian Ocean Tuna Commission. Available at: www.iotc.org/documents/report-presumed-iuu-fishing-activities-eez-somalia (accessed 20 January 2018).

Jensen, K. B. (2013) 'Child slavery and the fish processing industry in Bangladesh'. *Focus on Geography*, vol 56, no 2, pp. 54–65.

Kelly, A. (2016) 'Nestlé admits slavery in Thailand while fighting child labour lawsuit in Ivory Coast'. *The Guardian*, 1 February 2016, Available at: www.theguardian.com/sustainable-business/2016/feb/01/nestle-slavery-thailand-fighting-child-labour-lawsuit-ivory-coast (accessed 8 January 2017).

Larson, E. (2015) 'Costco sued over claims shrimp harvested with slave labor'. *Bloomberg.com*, 19 August 2015. Available at: www.bloomberg.com/news/articles/2015-08-19/costco-sued-over-claims-shrimp-is-harvested-with-slave-labor (accessed 8 January 2017).

Lawrence, F. and Hodal, K. (2017) 'Thailand accused of failing to stamp out murder and slavery in fishing industry'. *Guardian*, 30 March 2017. Available at: www.theguardian.com/global-development/2017/mar/30/thailand-failing-to-stamp-out-murder-slavery-fishing-industry-starvation-forced-labour-trafficking (accessed 3 October 2017).

Lawrence, F., McSweeney, E., Kelly A., Heywood M., Susman D., Kelly, C. and Domokos, J. (2015) 'Revealed: Trafficked migrant workers abused in Irish fishing industry'. *Guardian*, 2 November 2015. Available at: www.theguardian.com/global-development/2015/nov/02/revealed-trafficked-migrant-workers-abused-in-irish-fishing-industry (accessed 8 January 2017).

Lebel, L., Mungkungb, R., Gheewala, S. H. and Lebel, P. (2010) 'Innovation cycles, niches and sustainability in the shrimp aquaculture industry in Thailand'. *Environmental Science and Policy*, vol 13, no 4, pp. 291–302. doi:10.1016/j.envsci.2010.03.005.

Legacy Phuket Gazette (2014) 'US downgrades Thailand to be among worst human trafficking hubs'. News report, 21 June. Available at: www.phuketgazette.net/thailand-news/US-downgrades-Thailand-be-among-worst-human-trafficking-hubs (accessed 20 January 2018).

Lewis, S. G., Alifano, A., Boyle, M. and Mangel, M. (2017) 'Human rights and the sustainability of fisheries', in P. Levin and M. Poe (eds.), *Conservation for the Anthropocene Ocean: Interdisciplinary Science in Support of Nature and People*. London: Elsevier/Academic Press, pp. 379–396.

Lindley, J. (2016) *Somali Piracy: A Criminological Perspective*. London: Routledge.

Lindley, J. and Techera, E. J. (2017) 'Overcoming complexity in illegal, unregulated and unreported fishing to achieve effective regulatory pluralism'. *Marine Policy*, vol 81, pp. 71–79.

Lusk, W. (2016) 'Pirates of the Caribbean'. *Caribbean Maritime*, 29 October 2016, Available at: https://issuu.com/landmarine/docs/cm29/8 (accessed 5 April 2018).

Mendoza, M. and Mason, M. (2016) 'Hawaiian seafood caught by foreign crews confined on boats'. *Associated Press*, 8 September 2016. Available at: www.ap.org/explore/seafood-from-slaves/hawaiian-seafood-caught-foreign-crews-confined-boats.html (accessed 7 January 2018).

NEAFC (2018) 'IUU fishing and MCS effort'. North East Atlantic Fisheries Commission. Available at: www.neafc.org/mcs (accessed 20 January 2018).

Niue Treaty (2013) 'Niue Treaty on Cooperation in Fisheries Surveillance and Law Enforcement in the South Pacific Region'. Available at: www.ffa.int/system/files/Niue%20Treaty_0.pdf (accessed 20 January 2018).

Organization of the African Unity (1994) 'African Maritime Transport Charter' (adopted in Tunis, Tunisia, June 1994). Available at: https://au.int/sites/default/files/treaties/7776-treaty-0017_-_african_maritime_transport_charter_e.pdf (accessed 5 April 2018).

Organization of the African Unity (2018) 'List of countries which have signed, ratified/acceded to the African Charter on Maritime Security and Safety and Development in Africa (Lome Charter)' (updated 8 February 2018). Available at: https://au.int/sites/default/files/treaties/33128-sl-african_charter_on_maritime_security_and_safety_and_development_in_africa_lome_charter.pdf (accessed 5 April 2018).

Ratner, B. D., Åsgård, B. and Allison, E. H. (2014) 'Fishing for justice: Human rights, development, and fisheries sector reform'. *Global Environmental Change*, vol 27, no 1, pp. 120–130. doi:10.1016/j.gloenvcha.2014.05.006.

Rosenblatt, J. (2017) 'Costco defeats lawsuit over shrimp linked to Thai slave labor'. *Bloomberg.com*, 24 January 2017. Available at: www.bloomberg.com/news/articles/2017-01-25/costco-defeats-lawsuit-over-shrimp-linked-to-thai-slave-labor (accessed 16 February 2017).

Simmons, G. and Stringer, C. (2014) 'New Zealand's fisheries management system: Forced labour an ignored or overlooked dimension?'. *Marine Policy*, vol 50, part A, pp. 74–80. doi:10.1016/j.marpol.2014.05.013.

Stringer, C. and Simmons, G. (2015) 'Stepping through the looking glass: Researching slavery in New Zealand's fishing industry'. *Journal of Management Inquiry*, vol 24, no 3, p. 253. doi:10.1177/1056492614561228.

Stringer, C., Hughes, S., Whittaker, D. H., Haworth, N. and Simmons, G. (2016) 'Labour standards and regulation in global value chains: The case of the New Zealand fishing industry'. *Environment and Planning A*, vol 48, no 10, pp. 1910–1927. doi:10. 1177/0308518X16652397.

Surtees, R. (2013) 'Trapped at sea: Using the legal and regulatory framework to prevent and combat the trafficking of seafarers and fishers'. *Groningen Journal of International Law*, vol 1, no 2, pp. 91–151.

Sutton, T. and Siciliano, A. (2016) 'Seafood slavery: Human trafficking in the international fishing industry'. Center for American Progress, Washington, DC.

Telesetsky, A. (2014) 'Laundering fish in the global undercurrents: Illegal, unregulated and unreported fishing and transnational organised crime'. *Ecology Law Quarterly*, vol 41, pp. 939–997.

UNCLOS (1982) 'United Nations Convention on the Law of the Sea'. Available at: www.un.org/depts/los/convention_agreements/texts/unclos/unclos_e.pdf (accessed 20 January 2018).

United Nations Office on Drugs and Crime (2013) 'Transnational organized crime in eastern Africa: A threat assessment'. Available at: www.unodc.org/documents/data-and-analysis/Studies/TOC_East_Africa_2013.pdf (accessed 5 April 2018).

United Nations Secretary-General (2010) 'Secretary-General remarks to the General Assembly informal meeting on piracy'. Available at: www.un.org/sg/en/content/sg/statement/2010-05-14/secretary-general-remarks-general-assembly-informal-meeting-piracy (accessed 15 September 2017).

United States Department of State (2017) 'Human trafficking & migrant smuggling: Understanding the difference'. Available at: www.state.gov/j/tip/rls/fs/2017/272005.htm (accessed 5 April 2018).

Verité (2015) 'Recruitment practices and migrant labor conditions in Nestlé's Thai shrimp supply chain'. Available at: www.verite.org/wp-content/uploads/2016/11/NestleReport-ThaiShrimp_prepared-by-Verite.pdf (accessed 5 April 2018).

Walk Free (2017) 'Global Slavery Index 2016'. Walk Free, Perth. Available at: www.globalslaveryindex.org/download/ (accessed 5 April 2018).

Webster, D. G. (2015) *Beyond the Tragedy in Global Fisheries*. Cambridge, MA: MIT Press.

12 Impacts of (extreme) depth on life in the deep-sea

Shaun P. Collin, Lucille Chapuis and Nico K. Michiels

The deep-sea environment

The realm of the deep-sea is the least understood and most extreme environment on Earth, occupying over 99 per cent of all living space with an average depth of approximately 4,000 metres (Angel, 1997). Previously, it was thought that there was little or no life at these extreme depths but, as more sophisticated methods of sampling are developed (Figure 12.1), this vast region is now considered habitable space with 168 times more habitats than on land (Cohen, 1994). Species diversity is also high, and many animals rely on a range of eco-physiological adaptations for survival in the face of extreme gradients in hydrostatic pressure, temperature, light, water density, sound and currents. This chapter focuses on the impacts of these physical factors on the behaviour of a range of deep-sea animals (but predominantly fishes because they have received the most attention) throughout the water column, which is vertically stratified into zones.

The open ocean can be divided into several zones (see Figure 12.2): the epipelagic (surface to 200 m), the mesopelagic (200 m to 1,000 m), the bathypelagic (1,000 m to over 3–4,000 m), the abyssopelagic (over 4,000 m) and the hadal (6,000–11,000 m) zones. The epipelagic zone supports complex food webs and almost all life in the ocean via a continuous downward rain of organic material (Angel, 1997). The zone each species occupies is dependent on a range of physical and eco-physiological factors, which governs whether each species occupies midwater environments (pelagic), is restricted to living on the bottom (benthic) or near the bottom (benthopelagic) (Figure 12.2). The mesopelagic or twilight zone holds the greatest biomass and diversity of animal life found in the ocean; many animals have silver sides with dark upper bodies and counter-illuminating light organs underneath. Many others are black, red or transparent, which are all effective adaptations to camouflage their whereabouts in the dim, down-welling light environment. The bathypelagic zone is dark and the most biologically barren region on Earth, where both food and mates are scarce (Herring, 2002). Benthopelagic, abyssopelagic and benthic organisms are generally scavengers, feeding on the continuous rain of organic and inorganic matter (marine snow) that falls from above

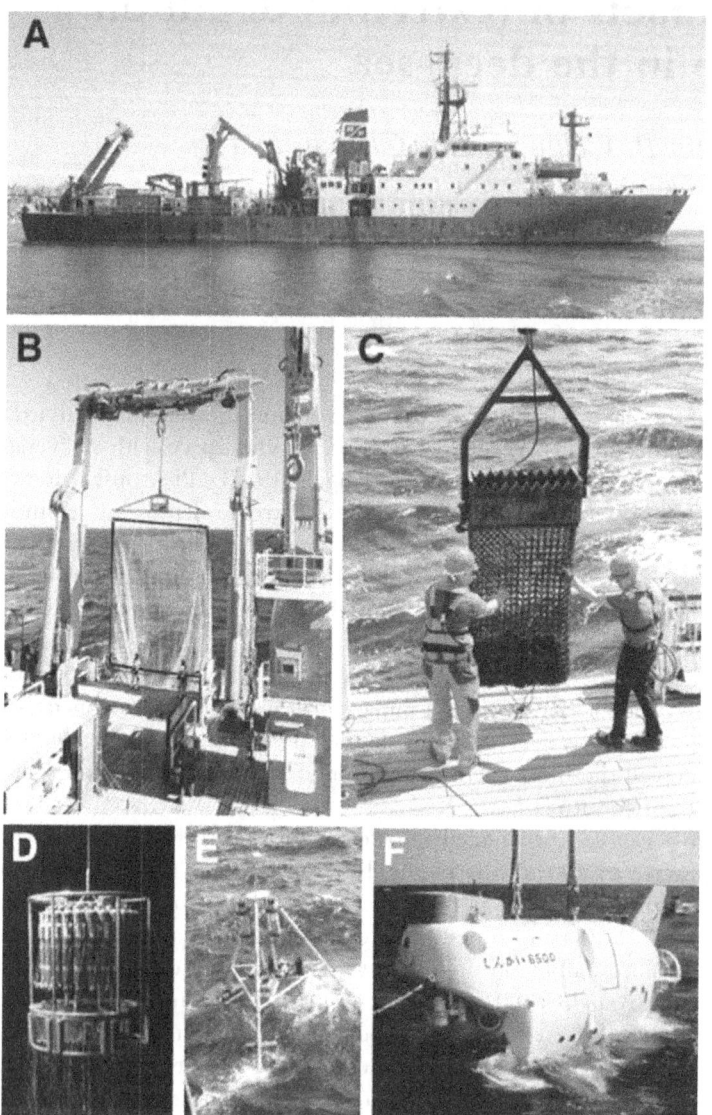

Figure 12.1 Series of photographs that illustrate the many ways of sampling the deep ocean.

Notes

A. R. F. Sonne in 2009 fitted with a stern-mounted A frame for deploying nets that can reach a depth of 2,000 m. B. Rectangular mesh trawling net deployed off the Sonne. Nets with closing cod ends are often used to sample animals within specific depth profiles. C. Bottom grab on deck following a deployment on a deep seamount. D. Conductivity, temperature and depth (CTD) instrument being retrieved. The rosette measures conductivity, temperature and pressure (depth). E. Hadal lander fitted with underwater cameras and bait traps used to attract and capture animals. F. Manned submersible, used to observe and film animals in their natural habitat.

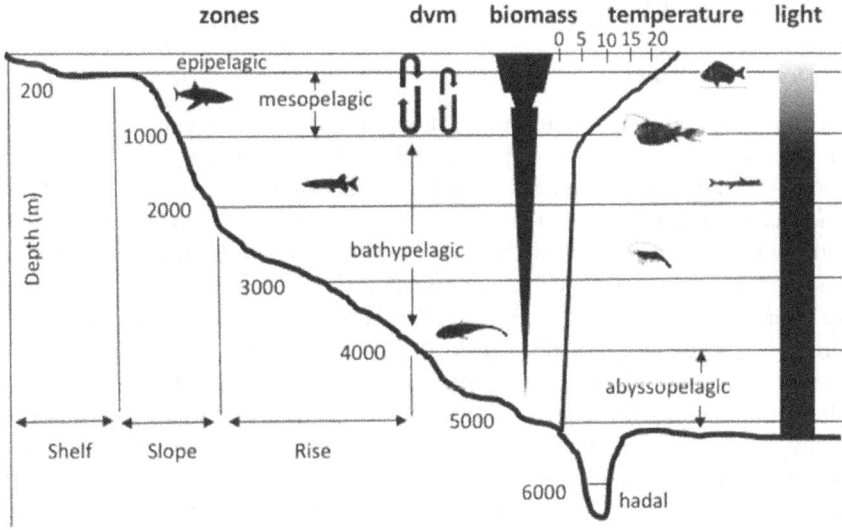

Figure 12.2 Schematic representation of the oceanic zones extending into the
deep-sea.

Source: adapted from Marshall (1979) and Angel (1997).

Notes
The limits of each zone are approximate and may differ across geographical regions and times of
the year. The topography of the continental shelf, slope and rise may also vary. The hadal zone
extends to depths in excess of 10,000 m in some regions. Temperature is in degrees Celsius.
DVM – diel vertical migrations.

(Shanks and Trent, 1980). Benthic inhabitants tend to be larger and stronger
swimmers than bathypelagic organisms due to a greater abundance of food
(Marshall, 1979). The biodiversity of organisms drops off radically in the
ocean trenches (hadal zones), which is predominantly occupied by amphipod
and decapod crustaceans, and rattail fishes (family Macrouridae, among the
most abundant of the deep-sea fishes) and snailfishes (family Liparidae with
elongated tadpole-like bodies). Like shallow water organisms, deep-sea
animals are uniquely adapted for survival within the constraints of their
environment. Some species are also able to migrate vertically through multiple zones following food resources such as the pelagic organisms that make
up the deep scattering layer (DSL), a horizontal zone within the water
column comprising high concentrations of marine organisms that reflect
sound waves generated by echo sounders (Afonso *et al.*, 2014).

The impacts of physical factors on eco-physiological processes and behaviour

Hydrostatic pressure

The pressures experienced from shallow water to the deepest parts of the ocean vary by a factor of approximately 1,000 (1 atm to over 1,000 atm). Despite this gradient in pressure, most animals can tolerate their surroundings and have evolved adaptations that permit relatively normal functioning, although biomass and the diversity of species drops dramatically within the hadal zone. Contrary to what many people think, deep-sea animals are able to tolerate the intense pressures at depth, partly because of the relative incompressibility of water and the high level of water content in their tissues (see Figure 12.3). Issues may arise with respect to animals migrating up and down the water column, but often the relative changes in pressures are minor (especially in the deeper-dwelling species), which allows daily oscillations over hundreds of metres. For vertical-migrating species of fish, which possess a swim bladder, gases are able to be absorbed or released to equilibrate the volume of the bladder, while maintaining neutral buoyancy. However, pressure also acts on all parts of the body and affects membrane transport mechanisms, biochemical reaction rates and other physiological processes.

Pressure can act on a range of tissue types, including cellular and subcellular membranes such as mitochondria, and affect enzyme kinetics and reaction rates. Interestingly, these effects can be partially compensated by increases in temperature (Sébert, 1993). Oxygen transport mechanisms (those that involve the transfer of oxygen from the surrounding water to cells), which vary from active processes like ventilation and circulatory convections to passive diffusion, are also affected by increases in pressure. Generally, the oxygen carrying capacity of deep-sea fishes is lower than that of shallow water fishes (Blaxter *et al.*, 1971; Graham *et al.*, 1985), despite sufficient levels of oxygen being available at depth. Therefore, in terms of energy production, deep-sea fishes may respire two times less than their shallow water counterparts, leading to lower metabolic rates and decreased levels of locomotion. However, more controlled (*in situ*) measurements are required to establish the true relationship between pressure and oxygen consumption, where lifestyle needs to be taken into account since active species of grenadiers (*Coryphanoides* sp.) have been found to respire 2–3 times faster than more sedentary species (*Sebastolobus* sp.) (Sébert, 1993).

There is also a direct relationship between muscle enzyme activity and pressure (depth) (Somero, 1991), especially in respect of those enzymes involved in aerobic and anaerobic pathways. Although enzyme kinetics are relatively unaffected by increases in pressure, enzyme activity is reduced with increasing depth. The white muscle of deep-sea fishes is also affected by increases in depth, resulting in a decrease in muscle fibre tension (strength), an increase in twitch contraction time and a reduction in the sensitivity of the

Figure 12.3 Range of invertebrate and vertebrate deep-sea animals that are known to have adaptations for living at extreme depths.

Notes

A. *Atolla* sp., jelly. B. *Abraiopsis* sp., squid. C. *Oneirodynia* sp., anglerfish. D. A deep-sea amphipod. E. *Argyropelecus* sp., hatchetfish. F. *Sigmops elongatus*, elongated bristlemouth G. *Sternoptyx diaphana*, diaphanous hatchetfish. H. *Chauliodus* sp., viperfish.

compound action potential of the motor nerves (Harper *et al.*, 1987; Wardle *et al.*, 1987). It appears that deep-sea fishes have sacrificed muscle strength and mobility in exchange for a lower metabolic rate (Swezey and Somero, 1982). However, most other homeostatic processes are maintained at depth, including general brain function and cellular and subcellular membrane fluidity, although buoyancy maintenance has an energy cost, where neutral buoyancy is achieved by reducing tissue density rather than via a swim bladder (Pelster, 1997).

As pressure increases, the partial pressures of the dissolved gases in the water decreases and their solubilities increase. This means that, as depth increases, the energy required for respiration (extraction of oxygen), the growth of skeletal muscle (the incorporation of carbon dioxide into calcium carbonate) and the inflation of the swim bladder (by a range of gases) increases substantially (Angel, 1997). Due to the fact that gases are more compressible than water, they become denser with depth and provide little buoyancy. This may explain why the swim bladder of most species of deep-sea fishes is either associated with large amounts of fat or lipid to reduce the overall density of the fish or it is lost and replaced by the deposition of lipids (such as triacylglycerol, wax esters and squalene) in the liver (Corner and Forster, 1969), bone tissue (Phleger and Laub, 1989) and extracellular lipid sacs (Pelster, 1997). In deep-sea sharks the liver may contribute up to 25 per cent of total body weight (Baldridge Jr, 1970; Van Vleet *et al.*, 1984).

Many deep-sea organisms have so-called watery tissues, where muscle tissue may be soft and possess a higher water content and the skeleton has reduced mineralisation (reduced concentration of calcium and phosphate ions). This increase in water content does not always translate to having neutral buoyancy, but it significantly reduces the weight of the fish in the water, thereby conserving energy. The head of the ophidiid, *Acanthonus armatus*, even possesses fluids of low osmolarity. Sodium and potassium concentrations in this fluid are lower than that of plasma, which contributes to buoyancy and helps maintain the head-up position of this fish in the water column (Horn *et al.*, 1978). Other methods of reducing the weight of the body include the incorporation of low density gelatinous material (glycosaminoglycans), as found in the larvae of many teleost species and in some species of sharks (Pelster, 1997). Long antennae in plankton, cilia in larval lancelets, well-developed pectoral fins, peduncular (tail) keels and heterocercal tails in many species of bony and cartilaginous fishes also provide hydrodynamic lift to achieve neutral buoyancy (Pelster, 1997).

Temperature

Marine temperatures range from 40°C in shallow tropical reef environments to −1.9°C (the freezing point of seawater) in polar regions; however, the average water temperature on Earth is between 3 and 4°C (see Figure 12.2) (Rice, 2000). This narrow range provides a cold, but stable, deep-sea environment across a large proportion of the planet. These cold temperatures can affect protein structure, as well as reduce membrane function and oxygen consumption (Sébert *et al.*, 1995), thereby limiting the biochemical kinetics of metabolism (Brown and Thatje, 2014). As described above, membrane fluidity may also be maintained by increased levels of lipids in deep-sea organisms and enzyme adaptation (Brown and Thatje, 2014). However, exceptions exist where super-heated water (between 300–400°C) percolates up through layers of sedimentary rock in the Earth's crust that underlies oceanic ridges, enters the water column through seeps in the substrate (hydrothermal vents) and is rapidly cooled producing 'smokers' as chemicals precipitate. Depending on the temperature, either black or white smokers erupt 30 m above the substrate and support a unique diversity of species, which obtain their energy from the chemicals streaming in through the process of chemosynthesis. Hydrothermal vent dwellers include bacteria, large bivalves (*Bathymodiolus* sp.), tube worms (*Riftia* sp.) and vent shrimps (*Bathograea* sp.). One species of 'eyeless' vent shrimp (*Rimicarus* sp.) occurs in high densities along the vent edge, feeding on the sulphide-loving bacteria that have accumulated over its body using specialised appendages. Instead of vision-forming eyes, members of this genus rely on sensory tissue beneath their transparent carapace to view their environment by detecting black body radiation (Pelli and Chamberlain, 1989; O'Neill *et al.*, 1995).

In the upper reaches of the water column (above 1,000 m), temperature is substantially more varied. The graded changes produce thermoclines, which dictate the depth range of many species due to physiological constraints on metabolic processes such as respiration, oxygen uptake and nerve conduction rates that convey sensory information to the central nervous system. These constraints have been found to influence planktivorous fish stocks in the Norwegian Sea, where mackerel were generally found in waters warmer than 8°C, while herring and blue whiting were mainly found in water masses between 2 and 8°C. In addition to potential differences in temperature tolerance, these groups were also spatially separated due to changing zooplankton densities (Utne *et al.*, 2012).

Light

The amount and spectral composition of sunlight underwater varies considerably with both water depth and proximity to land (Jerlov, 1976). Ocean water preferentially absorbs both short and long wavelengths of sunlight and transmits wavelengths maximally between 460 and 480 nm (Smith and Baker, 1981). Both scattering and absorption effects combine to reduce both the intensity and spectral bandwidth of sunlight available at depth. The brightly lit epipelagic zone supports a high biodiversity of animals with a full spectral range available for vision in the euryspectral zone between 0 and 15 m. Beyond this depth, wavelengths longer than 600 nm are almost completely gone and the remaining spectrum becomes narrower than what the organisms living in this zone are likely to be able to perceive (i.e. stenospectral zone) (Meadows *et al.*, 2014). The down-welling sunlight is further attenuated with increasing depth in the mesopelagic zone. Beyond 200 m, the spectral range of light is reduced to light at wavelengths around 480 nm (i.e. in the blue/ green part of the spectrum), which shows the least attenuation in the water column. At depths greater than 1,000 m, the number of photons originating from solar radiation is affectively reduced to zero (Warrant and Locket, 2004). Although often functionally reduced compared to those found in surface species, the eyes of deep-sea organisms, especially vertebrates, are well developed and show adaptations to reduced light levels, such as the development of larger eyes and an increase in the number and length of rod photoreceptors to maximise photon capture (Figure 12.4) (Locket, 1977; Collin, 1997). In many species, complex colour vision is lost in a trade-off between sensitivity and resolution, although the presence of cone photoreceptors in the eyes of some species (Collin *et al.*, 1998; Bozzano *et al.*, 2007; de Busserolles *et al.*, 2017) suggests that some form of colour vision may be retained to enhance the detection of bioluminescent (chemiluminescent) emissions. There is also a gradual transition from an extended visual scene to one that is dominated by point sources of bright bioluminescent flashes (Warrant and Locket, 2004), which has driven the evolution of a complex range of ocular specialisations (Collin and Partridge, 1996; Collin, 1997; Biagioni *et al.*, 2016).

Bioluminescence in the deep-sea is produced by either a luciferin–luciferase chemical reaction or a culture of bacteria that produces a bright emission, typically in the blue-cyan spectral range of 470–480 nm. It is used to camouflage the silhouette of animals seen from below by matching the intensity and spectral composition of the emissions of ventral light-producing organs or photophores to the down-welling sunlight penetrating to the upper reaches of the mesopelagic zone (Case *et al.*, 1977). The patterns of ventral photophore organs in many species of deep-sea bioluminescent sharks have recently been found to explain speciation in response to maintaining isoluminance as a counter illumination strategy to avoid predation from upward-looking predators (Claes *et al.*, 2014). Bioluminescence can also be used to find prey by either probing the area in front of the eyes with rostral-facing luminous organs (i.e. in the lanternfish, *Diaphus lucidus*) or by using light-emitting bacterial cultures housed in small sacs at the end of a lure suspended over a large mouth lined with needle-like teeth to attract prey (in the anglerfish, *Ceratius hobolli*) (Foran, 1991).

Other species such as the omnivorous gulper eel, *Eurypharynx pelicanoides*, has a large, gaping mouth that eats almost everything in its path, which is then stored in its oversized stomach before slowly digesting it (see Figure 12.4). Some species of female anglerfish (*Haplophryne mollis*) also attract mates using their bioluminescing genitalia; males become parasitic on the female, reduced to a sperm sac that provides the female with a way of producing young without having to attract further reproductive partners. Bioluminescent

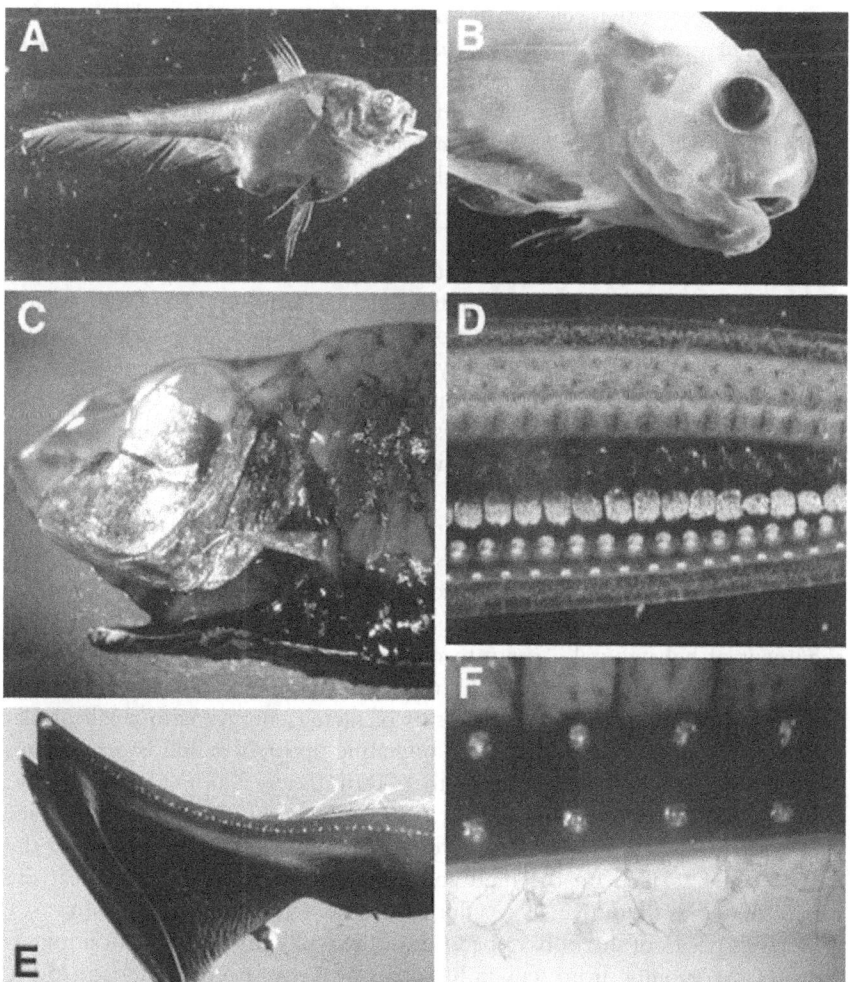

Figure 12.4 Examples of eco-physiological adaptations of deep-sea fishes to the deep-sea environment.

Notes

A. Macrourid rattail showing its long, tapering tail typical of many benthopelagic fishes. *B.* Liparid fish showing transparent tissue in the head. This group has very watery tissues to reduce body weight. *C.* Lateral view of the head of the barrel eye *Opisthoproctus* sp. showing its tubular eyes pointing upwards to view potential prey above against the background of downwelling sunlight. *D.* The body of the fangjaw *Gonostoma* sp. showing regular rows of reflecting mirror-like scales above several rows of ventrally-facing photophores. *E.* The gulper eel, *Eurypharynx pelicanoides*, showing its enormous mouth (here almost closed), which it uses to capture food. Note the rows of long hair cells (neuromasts) of the lateral line system along the dorsal margin of the body, sensitive receptors that detect minute hydrodynamic disturbances. *F.* Close up of the lateral flank of the viperfish *Chauliodus sloani* showing two rows of photophores above a thick layer of gelatinous material to increase buoyancy.

emissions can be used to stun or confuse potential predators and some shrimps, worms, squid and fishes release a luminous decoy to attract a predator's attention. Three genera of slingjaws (*Malacosteus*, *Pachystomius* and *Aristostomias*) have evolved a suborbital photophore that emits red bioluminescence to communicate with conspecifics to potentially attract mates (Douglas *et al.*, 1999). In flashlight fish (*Photoblepharon*, *Anomalops*), it has been suggested that this anatomical association between pupil and suborbital light organ is used for active detection of other species and conspecifics (Howland *et al.*, 1992; Hellinger *et al.*, 2017).

Although many species are restricted to quite narrow layers of the ocean based on a range of factors including food availability, eco-physiological responses to physical factors and sensory adaptations to avoid predation, many groups of animals migrate vertically up and down the water column in response to light. This diel vertical migration occurs in many species of plankton, shrimps and myctophid lanternfishes, where the circadian changes in the levels of sunlight, detected by sensitive eyes possessing over 85 million rod photoreceptors in *Myctophum brachygnathum* (de Busserolles *et al.*, 2014), initiate an upward migration towards the surface at dusk. The pineal organ (a small gland found in the brain of vertebrates) is also suspected to play a role in integrating light input over time in order to provide a mechanism for driving circadian activity patterns around vertical migration and photoentrainment or the alignment of behavioural activity to changes in illumination (Bowmaker and Wagner, 2004). The distance of the migration varies according to body size, and no doubt physiological mechanisms of saving energy, with small planktonic animals of only 1 mm in length migrating up to 20 m and larger shrimps of up to 6 cm in length migrating up to 1,000 m (Rice, 2000). The lanternfish, *Ceratoscopelus warmingii*, migrates over 1,500 m to within 100 m of the surface every night, only to return to its preferred depth of up to 1,800 m during the day, in order to feed on the high concentrations of phytoplankton near the surface. It avoids the surface layers during the day to avoid visual predators.

Several species of sharks have been also shown to exhibit diel vertical migrations (DVM) ranging from tens to hundreds of metres from the surface to the mesopelagic zone (200–1,000 m), or between the mesopelagic and bathypelagic (1,000–4,000 m) zones (Sims *et al.*, 2006; Kyne and Simpfendorfer, 2010). Pelagic sharks such as the blue shark, *Prionace glauca*, make regular DVMs during the day between the surface and depths of several hundred metres and are able to reach depths exceeding 600 m in around 10 minutes (Carey *et al.*, 1990). This means that their eyes have to adapt quickly to significant changes in light intensity in the order of about 7–8 log units (Warrant and Locket, 2004).

It has been suggested that in at least three groups of deep-dwelling elasmobranchs (urotrygonid stingrays, orectolobid wobbegongs sharks and scyliorhinid catsharks), fluorescence (photoluminescence) is used as a mechanism for species recognition by creating a greater luminous contrast against the surrounding background at depth (Sparks *et al.*, 2014; Gruber *et al.*, 2016; Bitton *et al.*, 2017). Fluorescence is the by-product of fluorescent compounds in the skin tissue that absorb the dominant high energy ambient blue light and re-emit it as longer, lower energy wavelengths resulting in patches of the body fluorescing in green, orange or red to the human visual system (Anthes *et al.*, 2016; Gruber *et al.*, 2016).

Density

The density of seawater is determined by three factors: hydrostatic pressure, temperature and salinity (the total quantity of dissolved salts in seawater, typically between 3.3 and 3.7 per cent). Depending on environmental conditions, this can result in vertically stratified layers. These layers allow some sensory cues such as odours and sounds to travel large distances due to the fact that little mixing occurs at the boundaries of these layers. Isopycnal layers (or regions of constant density) may allow chemical cues and sounds to travel thousands of kilometres. As described earlier, hydrostatic pressure is generally a function of depth or the weight of the overlying water per unit area. Although water is only very slightly compressible, it will increase in weight with depth at the same temperature. The vertical stratification of the water column is predominantly determined by the relative levels of evaporation of water at the sea surface (which would increase the density and salinity) and the levels of rainfall (which would increase the buoyancy of the upper wind-mixed surface layers), although riverine inputs of freshwater can also reduce the density of the upper regions of the water column (Angel, 1997). Following evaporation, the denser surface waters sink beneath (lighter) surface layers, at convergences, sliding down layers of equal density (pycnocline).

Sound

Water is a very good conductor of sound (travelling about 1,500 m/s compared to 300 m/s in air), with low frequencies travelling further than high frequencies which are quickly absorbed. In some regions, sounds of relatively low frequencies (around 10 kHz) may even penetrate the entire depth of the water column (thousands of metres below). The long-range propagation of sound depends on temperature, salinity and hydrostatic pressure. In regions where a temperature gradient exists, there is a characteristic decrease in the speed of sound with decreasing temperature. However, at a depth of approximately 750 m, the variations in temperature are so small that temperature becomes uniform (isothermal). From that point, the speed of the transmission of sound increases as depth (and therefore pressure) increases. At the interface of the thermocline and the isothermal depths, and where the speed of sound is at its lowest, an interface forms, which creates a deep sound channel that allows the transmission of low-frequency sound to propagate with little loss of energy over 25,000 km. This sound fixing and ranging (SOFAR) channel may allow many fishes and marine mammals to communicate over long distances (see Figure 12.5). A number of benthopelagic deep-sea fishes

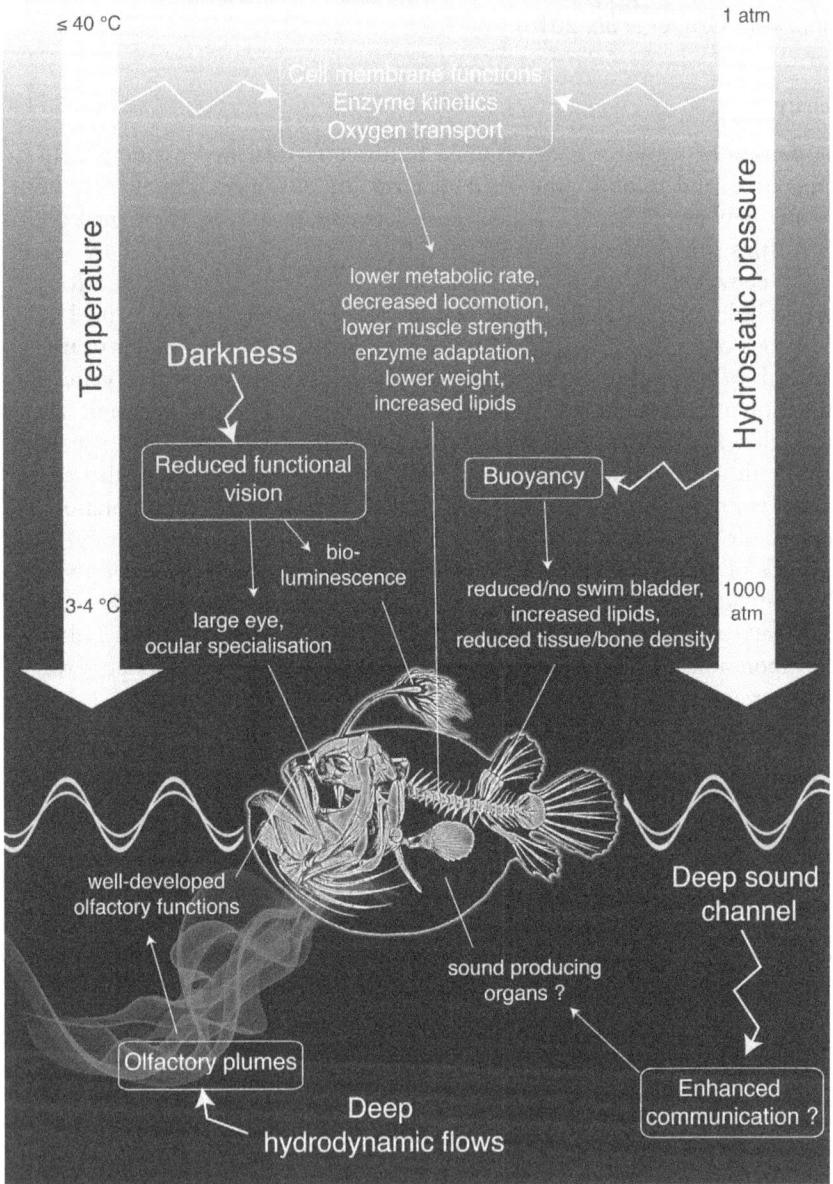

Figure 12.5 Summary diagram representing some of the important physical factors of the deep-sea.

Notes
The diagram shows the eco-physiological consequences of physical factors on deep-sea fishes, which have evolved numerous adaptations to live in these extreme conditions; atm – atmosphere.

(e.g. macrourids) have sound-producing muscles around their swim bladders (Marshall, 1967) which may be used to produce sound during reproduction (Mann and Jarvis, 2004).

Hydrodynamic changes

Surface currents are predominantly driven by winds, which produce a series of global gyres or circular currents, which can bring either warm or cold water. Smaller swirls and eddies may also be produced, often altering plant and animal growth in local regions. Deep circulation is driven by the sinking and rising of water of different density (produced by differences in salinity and temperature). In high latitudes, dense (cold saline) water sinks and is replaced by warmer, less saline water in lower latitudes giving rise to a series of interleaved 'pancakes' of water masses flowing slowly over and under one another in the deeper regions of the ocean. These complex systems are responsible for ocean mixing to help stabilise the water chemistry but also to carry oxygen to the deeper parts of the water column to support life.

At more local levels, water movement can aid in the transport of odours, which can travel long distances when aided by currents. Dead organic matter rains down from above, settles on the seafloor and forms olfactory trails or plumes, which are carried by bottom currents and detected by the olfactory system of fishes. Based on the relative size of the olfactory bulb, which receives input from the olfactory receptors in the nose, many species of benthopelagic fishes (i.e. macrourids, brotulids and synaphobranchids) are thought to be olfactory specialists (Marshall, 1979). Recognition of mates also occurs through chemical secretions or pheromonal communication, which can be carried horizontally in narrow 'pancakes' that can be less than 2 m in thickness but can carry reproductive cues for up to 1 km. As described earlier, the water column becomes stratified below the thermocline and complex layering occurs in response to changes in salinity (haloclines), density (pycnoclines) and currents. Based on average abundance, it has been suggested that mesopelagic hatchetfishes (*Argyropelecus hemigymnus*) are able to locate potential mates following the release of a pheromone into one of these horizontal transport systems in only one day instead of the typical eight days (Jumper and Baird, 1991).

Local hydrodynamic disturbances may also indicate the presence of food, especially at depths where vision is not well developed. Most animals are sensitive to both touch (proprioception) and minute water movements via the lateral lines system (see Figure 12.4). The sense of touch or the somatosensory system is well developed in tripod fish (*Bathypterois* sp.), which sit on the tips of their elongated pectoral (two) and caudal fins rheotactically oriented to take advantage of the incoming current to bring food. Many Antarctic fishes, which possess sensitive barbels extending from their lower jaw, also use these to probe the substrate in search of food (Montgomery and Pankhurst, 1997). The neuromasts of the lateral line system are groups of hair cells distributed either superficially (free) over the skin or within rigid or membranous canals beneath the skin that detect characteristics of the water movement (direction, acceleration and velocity) via the differential displacement of a gelatinous cupula covering. Some species increase the height of the cupula or raise the superficial neuromasts onto a small papilla (and therefore above the boundary layer) to enhance sensitivity to slow flows and low frequencies (see Figure 12.4) (Montgomery and Pankhurst, 1997). Deep-sea fishes such as *Poromitra* sp. possess lateral line canals that are wide and membranous and provide as much as a 100-fold increase in sensitivity in the 5–10 Hz range compared to shallow water fishes (Denton and Gray, 1988). The lateral line system typically operates over only one to two body lengths, but persistent vortices generated by the wake of passing animals can be detected at longer distances. Interestingly, the eel-like and tapering tails of many deep-sea fishes would not generate such hydrodynamic disturbances (when compared to the homocercal tails of many shallow water species) (see Figure 12.4), which could be considered a form of lateral line camouflage (Montgomery and Pankhurst, 1997).

Future opportunities for studying life in the deep-sea

The deep-sea has a critical role in the functioning and buffering of the Earth's systems by absorbing heat and calcium carbonate from the atmosphere, offering unique hotspots of biodiversity and biomass and providing numerous essential commercial resources and services. Its natural resources are not yet fully realised, but care and restraint must prevail if we are to protect these resources. As noted in Chapter 6, monitoring ocean health is critical, as is continued exploration of little known species and habitats. More sophisticated research vessels are now circumnavigating the globe probing the deep-sea in search of ways of understanding the geological origins of our planet, assessing the changing levels of biodiversity, the effects of climate change and the location of rich natural resources (see Figure 12.1). These ships are now able to deploy advanced instrumentation such as remotely-operated vehicles and manned submersibles to dive deeper and allow direct observations of the deepest regions in high definition (see Figure 12.1F). Robotics, satellite mapping, imaging and structural engineering enable manipulative experimentation and unprecedented prospects to fill the numerous knowledge gaps in our understanding of the deep-sea environment. All of these advances provide future opportunities to critically assess and conserve this vast ecosystem and possibly provide valuable benefits for the biomedical and other industries by way of bioprospecting (Thurber *et al.*, 2014). Extracting wealth from the deep-sea is an area of considerable interest in many national contexts, particularly for developing countries. It is often identified as an area for development within the context of the blue economy (Ramirez-Llodra *et al.*, 2011).

The deep-sea is not immune to the impacts of human activities. Sedimentary slopes, cold-water corals, canyon and seamount communities have all been described as the deep-sea habitats most at risk from the impacts of disposal, exploitation and climate change (Ramirez-Llodra *et al.*, 2011). Added to these risks are the pressures on species and habitats from extreme human behaviours, as explored in Chapter 11. As technological advances in exploring the ocean depths have been made, so has the ability to extract resources. Therefore, while deep-sea industrialisation expands steadily, so do the challenges faced by this ecosystem. Overfishing and the resultant declines in shallow water species are placing increased commercial fishing pressures on the fringes of the deep-sea environment (over 1,000 m, i.e. fisheries in orange roughy, grenadiers and alphonsino) (Clark *et al.*, 2015). Although expensive and logistically challenging, the likelihood of the oil and gas industry needing to tap into deeper deposits will continue to increase, despite the obvious advantages of renewable energy produced by wave, wind and solar technologies. The deep-sea could also represent an untapped source of valuable metals (i.e. nodules or crusts that sit on the seafloor, which are rich in manganese as well as nickel, cobalt and copper) when land-based sources are depleted. The International Seabed Authority has issued some exploration licences for seabed mining in areas beyond national jurisdiction, but only Papua New Guinea has authorised mining in the deep-sea within its jurisdiction to date (Folkersen *et al.*, 2018).

Although equally expensive, the deep-sea could also represent a potential sink for depositing harmful carbon dioxide from terrestrial-based activities, either as a liquefied gas or in frozen form (Rice, 2000), based on the fact that the ocean may not be able to continue to absorb the current levels of carbon dioxide. However, a more sustainable solution is to radically reduce our current levels of carbon dioxide emissions. Conservation of deep-sea habitats represents a difficult challenge considering the extreme environment, as well as large knowledge gaps that still remain on the functioning of its ecosystems, its biodiversity and its resilience to change (see Figure 12.5) (Levin and Le Bris, 2015). Given existing activities in the deep-sea, ecological restoration is another area where greater scientific research is needed (Van Dover *et al.*, 2014). Linking technological development with more fundamental knowledge on extreme deep-sea ecosystems, and establishing links between research, industry, management and policy, would help develop conservation and management options and sustainably achieve blue economy goals.

References

Afonso, P., McGinty, N., Graça, G., Fontes, J., Inácio, M., Totland, A. and Menezes, G. (2014) 'Vertical migrations of a deep-sea fish and its prey'. *PLoS ONE*, vol 9, no 5, p. e97884.

Angel, M. V. (1997) 'What is the deep sea?', in D. J Randall and A. P Farrell (eds.), *Fish Physiology*. London: Academic Press, pp. 1–41.

Anthes, N., Theobald, J., Gerlach, T., Meadows, M. G. and Michiels, N. K. (2016) 'Diversity and ecological correlates of red fluorescence in marine fishes'. *Frontiers in Ecology and Evolution*, vol 4, no 126, pp. 1–19.

Baldridge Jr, H. D. (1970) 'Sinking factors and average densities of Florida sharks as functions of liver buoyancy'. *Copeia*, vol 4, pp. 744–754.

Biagioni, L. M., Hunt, D. M. and Collin, S. P. (2016) 'Morphological characterization and topographic analysis of multiple photoreceptor types in the retinae of mesopelagic hatchetfishes with tubular eyes'. *Frontiers in Ecology and Evolution*, vol 4, no 25, pp. 1–13.

Bitton, P-P., Harant, U. K., Fritsch, R., Champ, C. M., Temple, S. E. and Michiels, N. K. (2017) 'Red fluorescence of the triplefin *Tripterygion delaisi* is increasingly visible against background light with increasing depth'. *Royal Society Open Science*, vol 4, no 3, p. 161009.

Blaxter, J. H. S., Wardle, C. S. and Roberts, B. L. (1971) 'Aspects of the circulatory physiology and muscle systems of deep-sea fish'. *Journal of the Marine Biological Association of the United Kingdom*, vol 51, no 4, pp. 991–1006.

Bowmaker, J. K. and Wagner, H.-J. (2004) 'Pineal organs of deep-sea fish: Photopigments and structure'. *Journal of Experimental Biology*, vol 207, pp. 2379–2387.

Bozzano, A., Pankhurst, P. M. and Sabatés, A. (2007) 'Early development of eye and retina in lanternfish larvae'. *Visual Neuroscience*, vol 24, no 3, pp. 423–436.

Brown, A. and Thatje, S. (2014) 'Explaining bathymetric diversity patterns in marine benthic invertebrates and demersal fishes: Physiological contributions to adaptation of life at depth'. *Biological Reviews*, vol 89, no 2, pp. 406–426.

Carey, F. G., Scharold, J. V. and Kalmijn, A. J. (1990) 'Movements of blue sharks (*Prionace glauca*) in depth and course'. *Marine Biology*, vol 106, no 3, pp. 329–342.

Case, J. F., Warner, J., Barnes, A. T. and Lowenstine, M. (1977) 'Bioluminescence of lantern fish (*Myctophidae*) in response to changes in light intensity'. *Nature*, vol 265, no 5590, p. 179.

Claes, J. M., Nilsson, D-E., Straube, N., Collin, S. P. and Mallefet, J. (2014) 'Isoluminance counterillumination drove bioluminescent shark radiation'. *Scientific Reports*, vol 4, p. 4328.

Clark, M. R., Althaus, F., Schlacher, T. A., Williams, A., Bowden, D. A. and Rowden, A. A. (2015) 'The impacts of deep-sea fisheries on benthic communities: A review'. *ICES Journal of Marine Science*, vol 73, suppl 1, pp. i51–i69.

Cohen, J. E. (1994) 'Marine and continental food webs: Three paradoxes?'. *Philosophical Transactions of the Royal Society B*, vol 343, no 1303, pp. 57–69.

Collin, S. P. (1997) 'Specialisations of the teleost visual system: Adaptive diversity from shallow-water to deep-sea'. *Acta Physiologica Scandinavica*, vol 161, pp. 5–24.

Collin, S. P. and Partridge, J. C. (1996) 'Retinal specializations in the eyes of deep-sea teleosts'. *Journal of Fish Biology*, vol 49, pp. 157–174.

Collin, S. P., Hoskins, R. V. and Partridge, J. C. (1998) 'Seven retinal specializations in the tubular eye of the deep-sea pearleye, *Scopelarchus michaelsarsi*: A case study in visual optimization'. *Brain, Behavior and Evolution*, vol 51, no 6, pp. 291–314.

Corner, E. D. S. and Forster, G. R. (1969) 'On the buoyancy of some deep-sea sharks'. *Proceedings of the Royal Society B*, vol 171, no 1025, pp. 415–429.

de Busserolles, F., Marshall, N. J. and Collin, S. P. (2014) 'The eyes of lanternfishes (*Myctophidae, Teleostei*): Novel ocular specializations for vision in dim light'. *Journal of Comparative Neurology*, vol 522, no 7, pp. 1618–1640.

de Busserolles, F., Cortesi, F., Helvik, J. V., Davies, W. I., Templin, R. M., Sullivan, R. K., Michell, C. T., Mountford, J. K., Collin, S. P. and Irigoien, X. (2017) 'Pushing the limits of photoreception in twilight conditions: The rod-like cone retina of the deep-sea pearlsides'. *Science Advances*, vol 3, no 11, p. 4709.

Denton, E. J. and Gray, J. A. B. (1988) 'Mechanical factors in the excitation of the lateral lines of fishes', in J. Atema, R. R. Fay, A. N. Popper and W. N. Tavolga (eds.), *Sensory Biology of Aquatic Animals*. New York: Springer, pp. 595–617.

Douglas, R. H., Partridge, J. C., Dulai, K. S., Hunt, D. M., Mullineaux, C. W. and Hynninen, P. H. (1999) 'Enhanced retinal longwave sensitivity using a chlorophyll-derived photosensitiser in *Malacosteus niger*, a deep-sea dragon fish with far red bioluminescence'. *Vision Research*, vol 39, no 17, pp. 2817–2832.

Folkersen, M. V., Fleming, C. M. and Hasan, S. (2018) 'The economic value of the deep sea: A systematic review and meta-analysis'. *Marine Policy*, vol 94, pp. 71–80.

Foran, D. (1991) 'Evidence of luminous bacterial symbionts in the light organs of myctophid and stomiiform fishes'. *Journal of Experimental Zoology Part A: Ecological Genetics and Physiology*, vol 259, no 1, pp. 1–8.

Graham, M. S., Haedrich, R. L. and Fletcher, G. L. (1985) 'Hematology of three deep-sea fishes: A reflection of low metabolic rates'. *Comparative Biochemistry and Physiology Part A: Physiology*, vol 80, no 1, pp. 79–84.

Gruber, D. F., Loew, E. R., Deheyn, D. D., Akkaynak, D., Gaffney, J. P., Smith, W. L., Davis, M. P., Stern, J. H., Pieribone, V. A. and Sparks, J. S. (2016) 'Biofluorescence in catsharks (*Scyliorhinidae*): Fundamental description and relevance for elasmobranch visual ecology'. *Scientific Reports*, vol 6, p. 24751.

Harper, A. A., Macdonald, A. G., Wardle, C. S. and Pennec, J-P. (1987) 'The pressure tolerance of deep sea fish axons: Results of challenger cruise 6B/85'. *Comparative Biochemistry and Physiology Part A: Physiology*, vol 88, no 4, pp. 647–653.

Hellinger, J., Jagers, P., Donner, M., Sutt, F., Mark, M. D., Senen, B., Tollrian, R. and Herlitze, S. (2017) 'The flashlight fish *Anomalops katoptron* uses bioluminescent light to detect prey in the dark'. *PLoS ONE*, vol 12, no 2, p. e0170489.

Herring, P. J. (2002). *The Biology of the Deep-Sea*. New York: Oxford University Press.

Horn, M. H., Grimes, P. W., Phleger, C. F. and McClanahan, L. L. (1978) 'Buoyancy function of the enlarged fluid-filled cranium in the deep-sea ophidiid fish *Acanthonus armatus*'. *Marine Biology*, vol 46, no 4, pp. 335–339.

Howland, H. C., Murphy, C. J. and McCosker, J. E. (1992) 'Detection of eyeshine by flashlight fishes of the family Anomalopidae'. *Vision Research*, vol 32, no 4, pp. 765–769.

Jerlov, N. G. (1976) *Marine Optics*. Amsterdam: Elsevier.

Jumper Jr, G. Y. and Baird, R. C. (1991) 'Location by olfaction: A model and application to the mating problem in the deep-sea hatchetfish *Argyropelecus hemigymnus*'. *The American Naturalist*, vol 138, no 6, pp. 1431–1458.

Kyne, P. M. and Simpfendorfer, C. A. (2010) 'Deepwater chondrichthyans', in J. C. Carrier, J. A. Musick and M. R. Heithaus (eds.), *Sharks and Their Relatives II. Biodiversity, Adaptive Physiology, and Conservation*. Boca Raton, FL: CRC Press, pp. 37–113.

Levin, L. A. and Le Bris, N. (2015) 'The deep ocean under climate change'. *Science*, vol 350, no 6262, pp. 766–768.

Locket, N. A. (1977) 'Adaptations to the deep-sea environment', in F. Crescitelli (ed.), *The Visual System in Vertebrates*. Berlin: Springer, pp. 67–192.

Mann, D. A. and Jarvis, S. M. (2004) 'Potential sound production by a deep-sea fish'. *Journal of the Acoustical Society of America*, vol 115, no 5, pp. 2331–2333.

Marshall, N. B. (1967) 'Sound-producing mechanisms and the biology of deep-sea fishes'. *Marine Bio-Acoustics*, vol 2, pp. 123–133.

Marshall, N. B. (1979) *Developments in Deep-Sea Biology*. Poole, UK: Blandford Press.

Meadows, M. G., Anthes, N., Dangelmayer, S., Alwany, M. A., Gerlach, T., Schulte, G., Sprenger, D., Theobald, J. and Michiels, N. K. (2014) 'Red fluorescence increases with depth in reef fishes, supporting a visual function, not UV protection'. *Proceedings of the Royal Society B*, vol 281, no 1790, p. 20141211.

Montgomery, J. and Pankhurst, N. W. (1997) 'Sensory physiology', in D. J. Randall and A. P. Farrell (eds.), *Deep-sea Fishes*. London: Academic Press, pp. 325–349.

O'Neill, P. J., Jinks, R. N., Herzog, E. D., Battelle, B-A., Kass, L., Renninger, G. H. and Chamberlain, S. C. (1995) 'The morphology of the dorsal eye of the hydrothermal vent shrimp, *Rimicaris exoculata*'. *Visual Neuroscience*, vol 12, no 5, pp. 861–875.

Pelli, D. G. and Chamberlain, S. C. (1989) 'The visibility of 350°C black-body radiation by the shrimp *Rimicaris exoculata* and man'. *Nature*, vol 337, no 6206, p. 460.

Pelster, B. (1997) 'Buoyancy at depth', in D. J. Randall and A. P. Farrell (eds.), *Fish Physiology*. London: Elsevier, pp. 195–237.

Phleger, C. F. and Laub, R. J. (1989) 'Skeletal fatty acids in fish from different depths off Jamaica'. *Comparative Biochemistry and Physiology Part B: Comparative Biochemistry*, vol 94, no 2, pp. 329–334.

Ramirez-Llodra, E., Tyler, P. A., Baker, M. C., Bergstad, O. A., Clark, M. R., Escobar, E., Levin, L. A., Menot, L., Rowden, A. A. and Smith, C. R. (2011) 'Man and the last great wilderness: Human impact on the deep sea'. *PLoS ONE*, vol 6, no 8, p. e22588.

Rice, T. (2000) *Deep Ocean*, London: Natural History Museum.

Sébert, P. (1993) 'Energy metabolism of fish under hydrostatic pressure: A review'. *Trends in Comparative Biochemistry & Physiology*, vol 1, pp. 289–317.

Sébert, P., Simon, B. and Barthélémy, L. (1995) 'Effects of temperature increase on oxygen consumption of yellow freshwater eels exposed to high hydrostatic pressure'. *Experimental Physiology*, vol 80, no 6, pp. 1039–1046.

Shanks, A. L. and Trent, J. D. (1980) 'Marine snow: Sinking rates and potential role in vertical flux'. *Deep Sea Research Part A. Oceanographic Research Papers*, vol 27, no 2, pp. 137–143.

Sims, D. W., Wearmouth, V. J., Southall, E. J., Hill, J. M., Moore, P., Rawlinson, K., Hutchinson, N., Budd, G. C., Righton, D. and Metcalfe, J. D. (2006) 'Hunt warm, rest cool: Bioenergetic strategy underlying diel vertical migration of a benthic shark'. *Journal of Animal Ecology*, vol 75, no 1, pp. 176–190.

Smith, R. C. and Baker, K. S. (1981) 'Optical properties of the clearest natural waters (200–800 nm)'. *Applied Optics*, vol 20, no 2, pp. 177–184.

Somero, G. N. (1991) 'Hydrostatic pressure and adaptations to the deep sea', in C. L. Prosser (ed.), *Environmental and Metabolic Animal Physiology*. New York: Wiley Liss, pp. 167–204.

Sparks, J. S., Schelly, R. C., Smith, W. L., Davis, M. P., Tchernov, D., Pieribone, V. A. and Gruber, D. F. (2014) 'The covert world of fish biofluorescence: A phylogenetically widespread and phenotypically variable phenomenon'. *PLoS ONE*, vol 9, no 1, p. e83259.

Swezey, R. R. and Somero, G. N. (1982) 'Skeletal muscle actin content is strongly conserved in fishes having different depths of distribution and capacities of locomotion'. *Marine Biology Letters*, vol 3, pp. 307–315.

Thurber, A. R., Sweetman, A. K., Narayanaswamy, B. E., Jones, D. O. B., Ingels, J. and Hansman, R. L. (2014) 'Ecosystem function and services provided by the deep sea'. *Biogeosciences*, vol 11, no 14, pp. 3941–3963.

Utne, K. R., Huse, G., Ottersen, G., Holst, J. C., Zabavnikov, V., Jacobsen, J. A., Óskarsson, G. J. and Nøttestad, L. (2012) 'Horizontal distribution and overlap of planktivorous fish stocks in the Norwegian Sea during summers 1995–2006'. *Marine Biology Research*, vol 8, no 5–6, pp. 420–441.

Van Dover, C., Aronson, J., Pendleton, L., Smith, S., Arnaud-Haond, S., Moreno-Mateos, D., Barbier, E., Billett, D., Bowers, K., Danovaro, R., Edwards, A., Kellert, S., Morato, T., Pollard, E., Rogers, A. and Warner, R. (2014) 'Ecological restoration in the deep sea: Desiderata'. *Marine Policy*, vol 44, pp. 98–106.

Van Vleet, E. S., Candileri, S., McNeillie, J., Reinhardt, S. B., Conkright, M. E. and Zwissler, A. (1984) 'Neutral lipid components of eleven species of Caribbean sharks'. *Comparative Biochemistry and Physiology Part B: Comparative Biochemistry*, vol 79, no 4, pp. 549–554.

Wardle, C. S., Tetteh-Lartey, N., Macdonald, A. G., Harper, A. A. and Pennec, J. P. (1987) 'The effect of pressure on the lateral swimming muscle of the European eel *Anguilla anguilla* and the deep sea eel *Histiobranchus bathybius*, results of Challenger cruise 6B/85'. *Comparative Biochemistry and Physiology Part A: Physiology*, vol 88, no 3, pp. 595–598.

Warrant, E. J. and Locket, N. A. (2004) 'Vision in the deep sea'. *Biological Reviews*, vol 79, no 3, pp. 671–712.

Part V
Synthesis

13 Addressing the challenges and harnessing the benefits of marine extremes

Erika J. Techera and Gundula Winter

Introduction

It is well-recognised that humans are dependent upon the ecosystem goods and services provided by the Earth's oceans. Yet in this book we have explored how humans have fundamentally altered the ocean environment and, in many ways, taken for granted the resources it provides with little regard for the future. The tide, however, has turned and increasing attention is being paid to our changing marine environment, the opportunities it provides, as well as the challenges we need to overcome to realise the full potential that the oceans may offer.

There are significant geographic areas, temporal changes and human behavioural issues about which we need new knowledge. There are also lessons to be learnt from past activities and events, as well as predictions we can make about the future based on existing data. It is clear that both theoretical and applied knowledge is rapidly being acquired in different oceanic contexts and focussed on a variety of marine extremes. Such knowledge must be communicated to and acted upon by decision-makers, industry and communities. This chapter synthesises the bodies of knowledge and case studies explored throughout this volume and identifies key research areas for the future.

Understanding marine extremes

In order to understand extreme environments, activities, events and impacts in the oceans fundamental scientific knowledge is needed. This knowledge can only be obtained through research. In some cases, such as the deep-sea, discoveries continue to be made in an environment about which little is known. In Chapter 12, Collin, Chapuis and Michiels explain what we know about the organisms and ecosystems at extreme depths, gained through a combination of technological developments that have allowed us to probe new depths and advance scientific expertise. We are beginning to understand what opportunities these genetic resources might hold for blue economy goals. Bio-prospecting is a growing area of interest because new medicinal and industrial products may be synthesised from novel species discovered in the deep-sea. At the same time, the authors have shown that human activities have impacted on this environment, despite its remoteness from coasts and communities. They remind us that human influences have reached the most extreme marine environments and that maintaining the health of the deep-sea is critically important.

Greater understanding is also needed in relation to the processes that affect coasts and estuaries. In Chapter 3, Winter, Hetzel, Peisheng, Hipsey, Mulligan and Hansen highlight the economic, social and environmental value of the coastal zone and the negative impacts of extreme events such as flooding, erosion and water quality disruptions. In the limited space available in the coastal zone, urban centres encroach on unique ecosystems with potential conflicts between economic development and environmental conservation. Managers need to integrate multiple uses in the coastal zone while building resilience to extreme events and climate change. Optimum management options to mitigate the effects of flooding, erosion and ecological disturbances caused by extreme weather events require advanced forecasting tools that oceanography and related disciplines must develop collaboratively.

The need for greater scientific understanding extends beyond geographic areas and processes to temporal changes in the oceans. In Chapter 6, Thomson, Cannell, Ghadouani, Fraser and Rayment explore the importance of monitoring ocean and estuarine health, and the definition of indicators that measure changes in marine environments. They present monitoring programmes at global and national scales, which not only provide critical foundational information for governments tasked with responding to change, but also inform the development of tools for modelling and forecasting future changes and impacts. Remote sensing technologies can now provide data from space, below water and along the coasts of unprecedented spatial and temporal coverage, which are not only of value to record environmental change but also to monitor illegal activities (discussed below).

Enhancing our understanding can include identifying new methodologies and approaches as well as uncovering new knowledge. In Chapter 4, Rogers and Burton explain the way community values can be quantified in monetary terms to be applied in decision-making processes. In this way, stakeholders can weigh up different options in a cost–benefit analysis framework. Rogers and Burton's focus is upon preferences for coastal hazard management where resources are limited. The authors discuss different approaches to capture these preferences and assign values so as to prioritise different investment options. There are powerful lessons to be learnt, for governments, communities and researchers, which may be transposed to other contexts.

In Chapter 9, Hepburn also explores the role of community knowledge in locally-led management interventions that increase environmental resilience towards marine extremes. Understanding how different actors can play a role in coastal stewardship is critically important as top-down responses to marine extremes cannot provide complete solutions. Good governance demands public participation and case studies such as the one outlined by Hepburn can help build capacity and understanding through a story of successful coastal custodianship over a significant period of time. Although specific approaches that work in one context will not necessarily be successful in another, valuable ideas may be generated by exploring case studies such as these.

Exploring human motivations – to protect our oceans or damage them – will also enhance our ability to respond to challenges. While Hepburn provides lessons on engaging communities in coastal custodianship, Lindley, Techera and Webster examine damaging human motivations in Chapter 11. The focus there is on drivers of crimes in marine areas and how they are interlinked or facilitated by certain factors. Extreme human behaviours, which here include maritime piracy, human trafficking, forced labour and illegal fishing, involve a range of actors with various motivations. Regulation alone cannot successfully address maritime crimes and it is critical to understand the underlying causes of human behaviour, why and how crimes are committed and what further interventions might be effective.

All of the chapters in this volume assist our understanding of marine extremes by broadening fundamental knowledge about people and the environment. A common thread running through these chapters is change. Research needs to be ongoing as the environment, communities and economy are constantly changing. Climate change is a particularly pressing issue but the chapters serve to highlight the breadth of change, including societal developments.

Impacts of marine extremes

The impacts of marine extremes are another common theme running throughout this volume. These not only include the impacts of natural processes on economic development and coastal communities, but also the effects of anthropogenic activities on the environment and social structures. Even natural extreme events are effected by humans, as anthropogenic climate change is shifting the baseline of what we consider 'extreme' today.

In Chapter 8, Cannell, Thomson, Schoepf, Pattiaratchi and Fraser highlight marine heatwaves to demonstrate the widespread impacts of just one aspect of the changing ocean environment. Past impacts can be measured and analysed but also utilised to help predict the future effects of extreme events. Winter *et al.* (Chapter 3) and Thomson *et al.* (Chapter 6) also highlight the importance of monitoring as well as forecasting and modelling the potential impacts of future extreme events. In these chapters the focus is on the impacts of natural events, albeit altered by climate change. It is clear, though, that human activities have resulted in damage to the environment. In Chapter 7, Snowball, Lehoux, Crawshaw, Savage, Hipsey, Ghadouani, McCulloch and Oldham draw attention to extreme pollution as an example of human activities that have left a lasting legacy for which both the environment and people are paying the price. The authors discuss land-based pollution that has dramatically increased with economic development, including industrial pollution, nutrient enriched stormwater and agricultural run-off. In Chapter 10, Pomeroy and Sequeira explore marine-based pollution and other impacts from aquaculture, but also highlight aquaculture's potential for sustainable blue economy development (discussed further below).

The impacts of illegitimate human behaviours are explored by Lindley *et al.* in Chapter 11. This illegal conduct has widespread ramifications, not just economic impacts, and the authors emphasise how it extends beyond environmental damage to human costs. Extreme human behaviours clearly have the potential to undermine the confidence and legitimacy of government and governance, and may unravel efforts to build community resilience to other marine extremes. Illegal activities add to the impacts on coasts and coastal communities from extreme events. Establishing a multi-disciplinary understanding of all aspects of marine extremes and their impacts is critical if effective and sustainable solutions are to be designed and implemented.

Responding to marine extremes

Responses to marine extremes feature strongly in many chapters in this volume. It is clear that technology is and will remain a critical factor in addressing many of the challenges explored in this book. In Chapter 12, for example, Collin *et al.* identify new technologies that have facilitated exploration of the deep-sea, overcoming the challenges of such an extreme environment. Illegal fishing is another issue which has benefited from the use of remote sensing technologies and satellite tracking for surveillance. Increasing computational capacity enables the development of better forecasting systems that can help keep coastal communities and economic activities safe.

There is little doubt that engineered measures to mitigate or prevent negative impacts from marine extremes are critical. But the chapters in this book have also highlighted the importance of integrated solutions that utilise nature's ecosystem services. In Chapter 5, Winter, Bryan and Ghisalberti focus squarely on this issue in the context of climate change and protecting our coasts. Aquaculture is another example where the challenge of food security is being addressed through a combination of natural processes and anthropocentric activities.

Responses to marine extremes must not be designed in isolation to the communities they seek to benefit. Legitimate ocean users and coastal communities must be involved in decision-making, whether this is related to industry's social licence to operate or government choices about climate change mitigation measures. In Chapter 4, Rogers and Burton emphasise the importance of capturing community preferences in respect of coastal hazard interventions. Community-led management of coastal areas and cooperation between people and government in building coastal resilience is also highlighted by Hepburn (Chapter 9). At the other end of the spectrum, the issue of maritime piracy provides a lens through which to explore multi-lateral, cooperative solutions to a global problem. Lindley *et al.* (Chapter 11) examine global responses to maritime piracy as another example to draw upon in designing future options for similar scale problems.

Challenges and benefits

The ocean has always presented challenges for humans; we cannot live in it but we depend on it. Natural environmental marine extremes, such as the deep-sea and extreme weather events, provide considerable practical difficulties for exploration, navigation and industrial operations. Yet commercial activities including shipping and energy production will continue to be pursued in the context of blue economy goals. Similarly, coasts will remain the principal places where people live. Coastal communities and economic goals are increasingly put at risk by climate change related challenges such as ocean warming, sea level rise and more frequent and intense extreme events.

The scope for future wealth from the oceans has also been recognised and articulated as blue economy policies and goals. There are opportunities, particularly for developing countries, to garner greater financial benefits that will assist their economic development and potentially provide economies and livelihoods to replace those that may become unsustainable in the future. In Chapter 10, Pomeroy and Sequeira highlight the prospects and challenges in relation to aquaculture: it can help resolve food security issues, provide jobs and economic opportunities, but it can also be a cause of extreme environmental impacts on our oceans. Aquaculture can be conducted further from shore to minimise coastal impacts, but it may then be affected by more extreme events triggered by climate change. The social implications extend beyond the provision of food. Without new employment opportunities there is a risk that as fisheries become depleted, and livelihoods are lost, people may turn to crime. Lindley *et al.* (Chapter 11) highlight how many Somali pirates were originally fishers who turned to illegal activities when local fish stocks became depleted. Therefore, blue economy goals can encourage new enterprises, but in pursuing these goals care must be taken to avoid unsustainable impacts on the oceans and people; lessons learnt from past activities must be considered.

Extreme human behaviours are problems created purely by people and therefore ought to be easier to address than unpredictable natural events. Yet addressing illegal activities in the ocean is problematic because of the difficulty of monitoring and surveillance. Here, new opportunities are presented by technologies such as remote sensing, satellites and drones that can illuminate activities in remote areas. Indeed, the above discussion draws sharply into focus the need for greater monitoring in general. The data collected from monitoring activities is valuable in a real-time sense, but also important for building long-term data sets to enable modelling, forecasting and prediction of future events. This will become increasingly important as extreme weather events, marine heatwaves for example, are expected to intensify and occur more frequently. As more data are collected and collated the importance of data management and analysis will continuously increase.

A research agenda for the future

Fundamental research into extreme environments including the deep-sea, as well as Polar and remote regions, is critical as long as these vast ocean areas remain unknown. The deep-sea in particular is already yielding resources of potential commercial and medical benefit to industry and human health. Fully comprehending these extreme environments, and the impacts of exploration, extraction and economic activities, is important to preserve them and their associated ecosystem services for future generations. Better understanding of oceanographic processes is essential in developing greater renewable energy opportunities from the ocean as well as offshore aquaculture. If managed well, such developments can provide economic and livelihood opportunities without compromising the environment.

Fundamental new knowledge is also needed about our changing ocean environment, including climate-induced changes. This requires baseline studies of different environments and activities, and thereafter longitudinal studies of change over time. A common theme flowing through many chapters in this book is a lack of data – or accurate and comprehensive data – in identified areas. It will be critical to fill these gaps if we are to accurately measure the ways in which the oceans are changing. This will be of value for better modelling and forecasting of future marine extreme events, but also for pre-empting and preventing catastrophic impacts.

As technology develops and becomes more affordable, large data sets are being produced. Over time, there will be an exponential growth in the data available regarding our oceans and our activities. Although big data has not been a focus of research in this volume, it is clearly a matter drawing considerable attention. Because much of the ocean environment is far from shore or deep in the water, remote sensing is the obvious choice to capture information that could not otherwise be obtained. Analysing this data through machine learning will be a critical area for research and application by scholars and practitioners.

Beyond the physical sciences, further social science research is needed in a number of areas. Sociological studies are becoming increasingly valuable as good governance practices are adopted which involve public participation and community engagement. How to work with diverse communities is critical knowledge for government and industry as will be methodologies to gauge social values and impact. This is most relevant when government is deciding what actions it will take to protect the environment and coastal communities from the extreme impacts of weather events and climate change, and whether it will approve activities which carry risks of extreme impacts. Shifting social values and connectedness to the changing ocean environment will become an important area as climate change effects become more apparent.

At the other end of the spectrum, a range of behavioural science research is needed to understand motivators and drivers of compliance and illegal behaviour so that human conduct can be modified. Illegal activities can have extreme effects on the ocean and dire human costs. The social impact of extreme negative behaviours must also be measured, perhaps quantified, and options identified to reduce damaging effects. Further economic scholarship is essential, particularly in applied areas, to enhance ways of capturing the cost of human activities and impacts on the ocean. Improved decision-making regimes are needed so that decisions can be made based on the full costs versus the true benefits, also accounting for non-monetary values.

Science communication remains a critical area in need of further attention. Whatever new discoveries are made and knowledge acquired, they must be embedded in law and policy designed to overcome marine extremes, ensure human safety and protect the oceanic environment. Clearly decision-makers must be made aware of scientific knowledge, but so too must communities and industry. Ways and means to better communicate are fundamental, and the role and value of social media in science communication is an area of growth worthy of further exploration to develop socially acceptable management strategies.

Social licence to operate is another area of research that will become increasingly important for industry that is involved in extreme environments or engaged in activities where the impact of an accident may be extreme even if the risk of it occurring is small. Increasingly, social scientists are exploring social licence in new ways including governments' social licence to regulate. Legitimacy of government intervention is being questioned, drawing attention to the need for further political science research.

It is clear that individual disciplines alone will not solve the problems. If marine extremes are truly to be understood and addressed, it will require crossing disciplinary boundaries. Considerable efforts can be seen in terms of multi-disciplinarity – bringing together bodies of knowledge from different fields – but genuine inter-disciplinarity is much harder to achieve and requires further efforts.

Conclusion

This volume has highlighted the need to better understand extreme ocean environments, events, activities and impacts; to communicate new knowledge; to involve all stakeholders; to identify ways to minimise and respond to impacts; and to harness opportunities and overcome challenges. The chapters in this volume make a valuable contribution to multi-disciplinary research across a spectrum of fields relevant to marine extremes. In doing so, this publication provides considerable hope that future collaborative solutions are not too far away.

Index

Page numbers in **bold** denote tables, those in *italics* denote figures.